# Michigan Gold & Silver

## Mining in the Upper Peninsula

by Daniel R. Fountain

Lake Superior
Port Cities Inc.

©2013 Daniel R. Fountain

All rights reserved. No part of this publication may be reproduced or transmitted in any form or by any means, electronic or mechanical, including photocopying, recording or any information storage and retrieval system, without permission in writing from the publisher.

First Edition: September 2013

Lake Superior Port Cities Inc.
P.O. Box 16417
Duluth, Minnesota 55816-0417 USA
888-BIG LAKE (888-244-5253)

5 4 3 2 1

Library of Congress Cataloging-in-Publication Data

Fountain, Daniel R.
   Michigan gold & silver : mining in the Upper Peninsula / by Daniel R. Fountain. – First edition.
      pages cm
   Includes bibliographical references and index.
   ISBN 978-1-938229-16-9
   1. Gold mines and mining – Michigan – Upper Peninsula – History. 2. Silver mines and mining – Michigan – Upper Peninsula – History. 3. Upper Peninsula (Mich.) – History. I. Title. II. Title: Michigan gold and silver.
   TN423.M46F69 2013
   622'.3422097749 – dc23                    2013028886

Editors: Konnie LeMay, Ann Possis, Bob Berg
Design: Tanya Bäck, cover; Amy Larsen, interior
Covers: Front cover, miners at the Kreig Gold Mine, photo courtesy Superior View; back cover, stamp mill drawing, photo from author's collection.
Printer: Sheridan Books Inc., Printed in the United States of America

# Dedication

This book is dedicated to
the memory of my father,
Robert G. Fountain, (1925-2012)
for starting me down this rocky road,
and for instilling in me his love of puns.

# Advance Praise for Michigan Gold & Silver

The story of gold and silver mining in Michigan's Upper Peninsula is invariably overshadowed by iron and copper, but the precious metals are the most fascinating mining of all.

*Michigan Gold & Silver* tells the powerful saga of the effort to extract "paying" quantities of the precious metals from a reluctant earth. Treasure is never easily surrendered. Powerful mining companies may litigate over iron and copper rights, but miners fought with guns, knives and raw fists over gold and silver claims!

While Dan Fountain's scholarship is outstanding, a treasure trove to serious students of mining, *Michigan Gold & Silver* doesn't skimp on the pure adventure of legendary lost mines and grizzled old prospectors coming to town with pockets bulging with heavy nuggets of valuable metal.

**– Frederick Stonehouse**
**Award-winning Michigan historian and author**

In *Michigan Gold & Silver*, author Daniel Fountain brings us a fascinating encounter with the history and lore of precious-metal exploration in Michigan's Upper Peninsula. A sequel to his 1992 book *Michigan Gold*, Fountain's new work contains expanded and updated sections on gold, as well new chapters on Michigan's 19th century silver rushes, plus an epilogue on recent discoveries of precious and base metals and the ongoing efforts to develop these deposits into new 21st century mines. Though focused primarily on history, Fountain does equal justice to the mineralogical science and distills that oft-complex language into a friendly and easy read. Destined to become a classic of Upper Michigan literature, this book is a must-have for history buffs, geologists and modern-day treasure seekers alike.

**– Shawn Carlson**
**Michigan mineralogist and author**

# Foreword

In the 20 years since *Michigan Gold, Mining in the Upper Peninsula* came out in 1992, it has sold more than 9,000 copies in five printings. Since the publication, I've been invited to give numerous presentations on gold mining history. Often in the question and answer sessions after the shows, I would get questions from people looking for more history: "What's that old shaft behind Teal Lake?" "What did they mine at the Holyoke?" "What about the old silver lead shafts?"

I began to appreciate that silver and gold were inextricably linked in Michigan's mining history. Although in the original book I mentioned silver as a byproduct of the gold mining, I felt that I hadn't done justice to silver mining's own rich history. *Michigan Gold & Silver, Mining in the Upper Peninsula* is my attempt to bring to light this facet of Michigan's mining history.

Ironically, despite two silver mining rushes – in Marquette County during the 1860s and in Ontonagon County in the 1870s – Michigan's only significant production of silver has come from the copper mines of Keweenaw, Houghton and Ontonagon counties. While the history of these mines is fascinating in its own right, it is beyond the scope of this book to recount the production of silver from the copper range. I've focused on the prospecting and mining ventures specifically targeting silver.

I have also taken this opportunity to expand and update my coverage of the gold mines of Michigan. I've added a few prospects that weren't covered in the original book, and I've learned more details about some that were previously covered.

I've tried to include as much detail as possible without encouraging trespass, as I realized that some readers will not be interested just in reading the history of the prospects but may want to explore the gold and silver ranges themselves. Not everyone who reads this book will be familiar with mining terms, and so I encourage readers to consult the glossary at the back of the book.

The prospectors of the previous centuries were amazingly thorough, as evidenced by the number of prospect pits and shafts found in even the most remote corner of the Upper Peninsula. Still, new discoveries of precious metals have been made in the late 20th and 21st centuries. Even the technology of prospecting for precious metals has matured over the last 20 years, and I try to contrast modern prospecting techniques with the old-timers' methods as well as introduce the latest precious mineral prospects in a new chapter.

The researcher's job, too, has changed since 1992. The vast amount of data available online, and particularly the search engines that enable us to find it, have made it possible to learn obscure and often unique information. I still ended up doing some research the old-fashioned way, poring over fragile 140-year-old newspapers and blurry microfilm copies to learn much of the history of some of the lesser-known operations.

I'm looking forward to the next 20 years and seeing what the technology and culture of the 21st century will bring to the search for gold and silver.

Dan Fountain
September 2013

*Foreword from the original* Michigan Gold

Ever since I was a child growing up in Ishpeming, Michigan, I had heard of the old Ropes Gold Mine, but never knew much about it. I shared a common misconception about gold mines, namely that they generally produced large nuggets of gold that anyone could find just by digging a bit. A couple of trips to the old mine's rock dumps and an ill-fated panning expedition to Gold Mine Creek (armed with a tin pie plate and plastic bags for my gold dust and nuggets) quickly dispelled these ideas, and I dismissed the Ropes as worthless.

When in the 1970s I began to hear that mining companies were surveying the area around the mine, I began to search out the history of the Ropes and Michigan Gold mines, which, as far as I knew, were the only mines in the area. As I dug into the available resources, I learned to my surprise that gold prospects were spread throughout Marquette County and across the Upper Peninsula. This book is my attempt to compile the stories of these mines and miners.

As I started to search through the local libraries for information on the mines, I immediately found that no comprehensive recounting of the many gold prospects existed. Contemporary reports issued by the Michigan Geological Survey and the Commissioner of Mines and Mineral Statistics mentioned many of the major prospects; the files in the John M. Longyear Research Library at the Marquette County Historical Society told of others. Back issues of the *Engineering and Mining Journal* occasionally featured reports on the new mines and companies and listed stock prices.

By far the richest source of information was the *Iron Ore*, Ishpeming's weekly newspaper from 1881 through 1953. George Newett, the paper's publisher, was a tireless backer of the city and especially its mining industries, both iron and gold. Nearly every new mining company was given a full write-up in his paper, including the location and extent of the exploration, the officers and financial arrangement of the company and, most important, how the reader could purchase stock in this promising young company! Newett was not one to pull his punches, however, being just as ready to condemn a mining company's management at the first sign of a swindle.

One disadvantage of relying on the press for historical information is the lack of any negative news about the prospects. As long as the gold veins held out, the promising company was news; but when the veins pinched or assays declined in value, the prospects were forgotten. By the same token, assay values published in the media must be read in context. Many prospects produced samples that held, by assay, several ounces of precious metal per ton of ore, yet were quickly abandoned. What was not reported was that it was not tons, but merely a few pounds or only ounces of the rich ore, were available to be mined!

This book is intended as a history, rather than a prospector's guide. Although legal descriptions are given identifying the location of the land held by the early explorers, the prospects themselves are not pinpointed. There are a couple of reasons for this. First, most of the prospects mentioned in this book lie on private property, and the owners do not always welcome visitors. Michigan trespass law requires the permission of the landowner before entering private land. Second is the matter of safety. Many of the shafts and adits dug a century ago are still open, lying hidden and forgotten in the woods. In some cases, water-filled shafts more than 100 feet deep are left unfenced, surrounded by mossy banks and piles of loose rock. Another reason I have not given the exact locations of some prospects is that they have (so far) eluded me!

Dan Fountain<br>September 1992

# Acknowledgments

The old-time prospector is often portrayed as a loner with only his burro as companion, but this modern literary "sourdough" could not have written this book without the contributions of many others.

The author extends his sincere thanks to:

Jim Rouse and Bruce Cox for sharing their encyclopedic knowledge of the Gogebic Range;

Gene Londo for showing me the Ropes;

Fred Stonehouse for his encouragement;

Keith Voss for his exhaustive research on the Michigan Gold Mine;

Steve Lehto for making me check my facts and hedge my bets;

Jack Steve for a great story;

Randy Leppala and the Rainbow Exploration Company;

Larry Roncaglione for his continued interest and input;

Kay Wallace Porter for her priceless photos and family history;

Fred Rydholm for stories and maps;

Shawn Carlson for inside information on obscure prospects;

Tom Quigley of Aquila Resources for setting me straight on some important details;

Rich Lassin for some little-known stories;

Lee DeGood for sharing his outstanding collection of mining stock certificates;

Marguerite Grummett Bergdahl, the last of the Grummetts;

Dick Nicholas and Ethel Valenzio for family photos;

Jack Deo of Superior View for sharing his collection of rare photos;

Rosemary Michelin, Linda Panian and Rachel Crary of the Marquette County Historical Society;

Willard Bodwell of Resource Exploration Incorporated for maps and modern history;

Fred Peterson of Longyear Realty Corporation for information on 20th century prospecting;

John Norby of Callahan Mining Corporation for cluing me in to some obscure locations;

the staff of the Carnegie Public Library, Ishpeming;

the staff of the Peter White Public Library, Marquette;

the staff of the Negaunee Historical Society;

the staff of the Iron County Historical Society, Hurley, Wisconsin;

Kurt Fosburg, my enthusiastic prospecting partner;

and especially to my wife, Judy, for the years of encouragement, suggestions, editing, love and moral support.

# Table of Contents

Foreword . . . . . . . . . . . . . . . . . . . . . . . . . . . . . . . . . . v
Acknowledgments . . . . . . . . . . . . . . . . . . . . . . . . . . . . ix
Upper Peninsula maps . . . . . . . . . . . . . . . . . . . . . . . . xii
Introduction . . . . . . . . . . . . . . . . . . . . . . . . . . . . . . . xiv
Timeline for Michigan Gold & Silver . . . . . . . . . . . xvii

| | | |
|---|---|---|
| Chapter 1 | Early Gold & Silver Discoveries. . . . . . . . . . . . . . . . . . . 2 |
| Chapter 2 | The 1860s Silver Lead Rush. . . . . . . . . . . . . . . . . . . . 12 |
| Chapter 3 | The Ontonagon County Silver Rush . . . . . . . . . . . . . 36 |
| Chapter 4 | The Ropes Gold Mine . . . . . . . . . . . . . . . . . . . . . . . . 60 |
| Chapter 5 | The Michigan Gold Mine . . . . . . . . . . . . . . . . . . . . . 96 |
| Chapter 6 | Michigan Range Prospects . . . . . . . . . . . . . . . . . . . . 109 |
| Chapter 7 | Ropes Range Prospects. . . . . . . . . . . . . . . . . . . . . . . 124 |
| Chapter 8 | The Dead River Gold Range . . . . . . . . . . . . . . . . . . 137 |
| Chapter 9 | Other Marquette County Prospects . . . . . . . . . . . . . 146 |
| Chapter 10 | Baraga, Iron & Dickinson County Prospects . . . . . . 166 |
| Chapter 11 | Gogebic Range Gold & Silver Mines . . . . . . . . . . . . 175 |
| Chapter 12 | Placer Mining . . . . . . . . . . . . . . . . . . . . . . . . . . . . . 191 |
| Chapter 13 | Fables & Frauds . . . . . . . . . . . . . . . . . . . . . . . . . . . 196 |
| Chapter 14 | Epilogue: 20th & 21st Century Gold Explorations . 202 |

Glossary. . . . . . . . . . . . . . . . . . . . . . . . . . . . . . . . . 214
Bibliography . . . . . . . . . . . . . . . . . . . . . . . . . . . . . 217
Index . . . . . . . . . . . . . . . . . . . . . . . . . . . . . . . . . . 232
Index of Graphics. . . . . . . . . . . . . . . . . . . . . . . . . 240
About the Author. . . . . . . . . . . . . . . . . . . . . . . . . 242

# Upper Peninsula - West

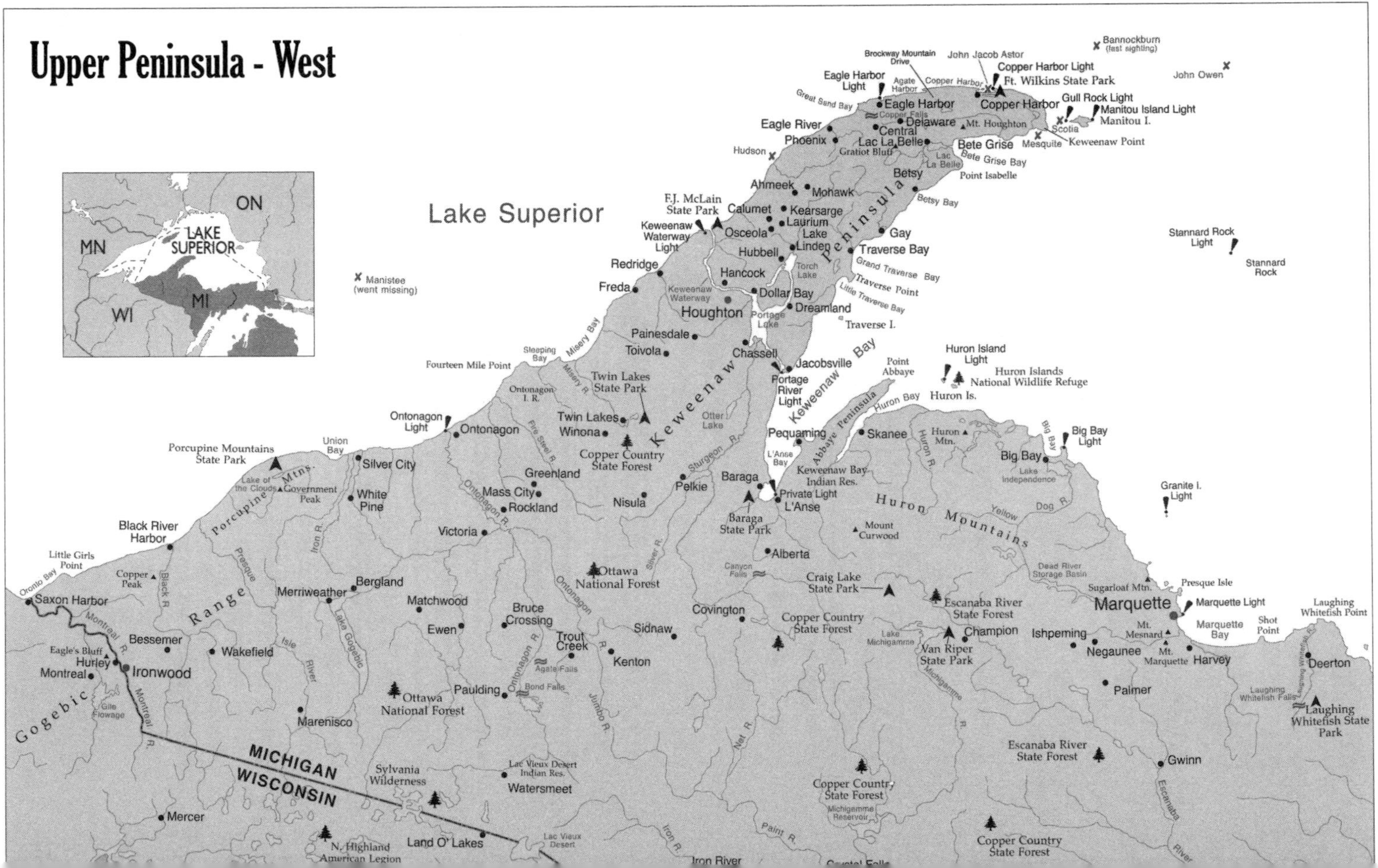

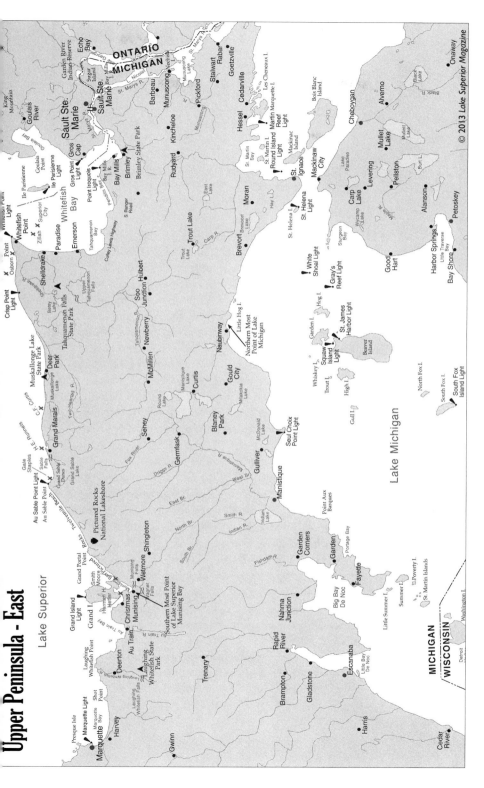

# Introduction

Michigan's Upper Peninsula is an ancient, rugged land. The contorted surface of its western half stands in mute witness to its ancient fiery birth. This part of the peninsula is made up of some of the oldest rock in the world – the more than 3 billion-year-old Canadian Shield. Predating the earliest life on earth, this ancient formation hosts an array of valuable and not-so-valuable minerals, some with familiar names like copper and iron ore, others with more exotic names of palladium, pyrrhotite and chalcopyrite.

Copper, the best known of the minerals found in the U.P. Copper Country and on Isle Royale, was mined thousands of years ago by prehistoric peoples. Mere centuries ago, early fur traders and missionaries to the area related stories about massive pieces of the virgin metal that could be dug from the ground. A few mining ventures were begun back in the 1700s and, after 1841, when Douglass Houghton, Michigan's first state geologist, documented the rich copper deposits of the Keweenaw Peninsula, a full copper rush to the Lake Superior region began. Mining operations got under way in earnest in 1845 with the opening of the Cliff mine near Eagle River.

Iron, too, became a U.P. economic driver. In 1845, as the Cliff copper mine was getting under way, the rich iron deposits of the Marquette Iron Range were discovered near present-day Negaunee. A party of surveyors under William Austin Burt was running a range line, the north-south line between two townships, when they noticed their compass making dramatic deviations from true north. A search of the surrounding area turned up specimens of high-grade iron ore. Within a few years, the towns of Negaunee and Ishpeming were founded and the iron ore mining industry of the Upper Peninsula was begun.

But the Upper Peninsula has another mineral legacy less known and with a more variable economic and social history. It has left, though, clues to its past in the familiar names found in the region: Silver City, Silver Mountain, Gold Mine Lake, Gold Mine Creek, Silver Lake, Silver Mine Lake and Silver Lead Mine Lake among them.

Those sparkling metals – the ones known to have sparked the famed "rushes" of the West – were also found in Michigan's U.P. In fact, the local Native people, familiar with the copper that they used for tools and jewelry, had also mined silver, though they knew nothing of the yellow metal, the gold, that so intrigued some of the earliest European explorers to the Lake Superior country.

Although most of those early explorers, and the later prospectors to follow, soon turned their attention to the more abundant copper and iron deposits, gold and silver would get their following – and would later bring their own gold and silver rushes to the Upper Peninsula. As early as 1845, though, small amounts of gold were found, and silver was a valuable byproduct of the Keweenaw copper mines.

Silver prospecting got the first foothold with the discovery of silver lead – a lead ore carrying silver – started a prospecting rush in northern Marquette County in the mid-1860s, though no successful mines were opened. The discovery of silver in Ontonagon County in the 1870s also generated interest and mining operations, but economic pressures and unfavorable geologic conditions condemned these mines to failure, despite an extensive formation of rich silver ore.

Gold, too, brought mostly unfulfilled hopes of great wealth. The Ropes Gold Mine, established in the early 1880s, ran for 14 years and produced $645,792 in gold and silver, but was never able to pay a dividend to its stockholders. Ropes did produce another dividend – more than 75 other gold mines and prospects were started in the U.P. in the months and years after its establishment.

The most famous of these was the Michigan Gold Mine, 2½ miles southwest of the Ropes, which produced spectacular specimens of free gold – so spectacular that one mine worker departing for Europe was discovered to have more than $2,000 worth of gold in his trunk "secured from this property when the eyes of the bosses were not upon him."

Most prospecting had ceased by the early 1900s, but the Great Depression and the increase in the price of gold from $20.67 to $35 per ounce brought about another prospecting boom in the 1930s. World War II called a halt to gold mining, and the prospects lay idle until abandonment of the gold standard in 1971. Once again, the resultant dramatic rise in the price of gold, along with improved metallurgical methods, attracted a $20 million redevelopment project to the Ropes mine, which again began producing gold in the fall of 1985. The reopened mine produced until 1989, when a combination of low gold prices, poor ore grade and a collapse of rock in the production shaft prompted its shutdown. Times change, though, and gold prices again have risen.

This book, then, is the story of those lesser known parts of the U.P.'s rich mineral history – and the final chapter on mining these minerals in the Upper Peninsula may yet remain to be written.

# A Timeline for Michigan Gold & Silver

| | |
|---|---|
| 1636 | "Mines of copper" on Great Lakes mentioned in Paris. |
| 1738 | Sieur de la Ronde explores Ontonagon area for copper, finds signs of silver. |
| 1770 | Alexander Henry searches for gold on the "island of yellow sands" on Lake Superior. |
| 1844 | Marquette iron range discovered. |
| 1845 | Douglass Houghton reportedly finds gold near Negaunee. |
| 1845 | New York and Lake Superior Mining Company mines for copper, lead and silver at Marquette. |
| 1845-6 | National Mining Company mines for silver at Silver Mountain. |
| 1846 | Jacob Houghton finds gold in copper ore in U.P. |
| 1847 | Gogebic Iron Range discovered. |
| 1854-5 | Silas C. Smith discovers gold in the city of Marquette. |
| 1855 | Austin Corser discovers silver-bearing quartzite in the Little Iron River, west of Ontonagon. |
| 1863 | Silas C. Smith finds argentiferous galena near Silver Lake, starting the silver lead rush. |
| 1864 | Gold found on the silver lead range. |
| 1865 | Waterman Palmer finds gold-bearing quartz vein near Palmer. |
| 1868 | Holyoke Mining Company ceases operations, signaling the end of the silver lead era. |
| 1872 | Austin Corser reveals his silver find, starting the Iron River silver boom. |
| 1875 | Iron River Silver Reduction Works built to process ore from the silver mines. |

The 1887 mill at the Ropes Gold Mine.
AUTHOR'S COLLECTION

| | |
|---|---|
| 1876 | The Cleveland Silver Mining Company closes its mine, ending the Iron River silver boom. |
| 1879 | Julius Ropes finds small amounts of gold and silver north of Ishpeming. |
| 1879 | Argentiferous galena found near Lake Gogebic. |
| 1881 | Ropes discovers commercial-grade gold ore and incorporates the Ropes Gold and Silver Company. |
| 1883 | Stamp mill installed at Ropes Gold Mine. First ore is processed, producing 40 ounces of gold. Curry shaft begun. |
| 1884 | 25-stamp mill installed in new building at Ropes. |
| 1885 | Gold discovered west of Ropes on Lake Superior Iron Company land. |
| 1886 | Power drills introduced at Ropes, replacing hand drilling. |
| 1887 | Gold worth $44,000 per ton discovered at Lake Superior Iron Company shaft. Michigan Gold Mine opened. |

| | |
|---|---|
| 1887 | New mill with 20 stamps erected at Ropes; expanded to 40 stamps in 1889. |
| 1888 | Gold ore worth $100,000 per ton found at Michigan mine. |
| 1890 | First underground diamond-drilling exploration on the gold range at Ropes. |
| 1891 | Fire Centre Mining Company incorporated to work gold veins discovered by Julius Ropes north of the Dead River. |
| 1892 | Curry shaft at Ropes mine reaches its final depth of 813 feet. |
| 1893 | The panic of 1893 leads to the closing of the Fire Centre properties. |
| 1896 | Michigan mine shuts down. |
| 1897 | Miners sue the Ropes Gold and Silver Company for back wages, forcing the mine to close. |
| 1899 | Ropes mine sold to Corrigan, McKinney & Company. |
| 1900-1 | Ropes mine tailings reprocessed with cyanide leaching, recovering $54,682 in gold and silver. |
| 1903 | Michigan mine reopened by the Tribullion Mining, Smelting and Development Company; shut down the same year. |
| 1905 | Michigan mine reopened as a silica mine by Ishpeming Gold Mines Company; later operated by Michigan Quartz Silica Company. |
| 1927 | Bjork & Lundin lease Ropes property and recover $1,000 worth of amalgam from old mill site. |
| 1932 | Ishpeming Gold Mining Company forms, recovers a small amount of gold by cyanide treatment of Ropes mine tailings. |
| 1933-4 | Michigan Gold Mines Inc. reopens Michigan mine with a new mill and produces 58 ounces of gold. |
| 1934 | Gold price rises to $35 per ounce. Calumet & Hecla buys Ropes Gold Mine, dewaters it and conducts an evaluation program, finding 1.5 million tons of gold ore worth $4.90 per ton. |
| 1937 | Michigan Gold Mining Company produces 51 ounces of gold from the Michigan mine. |
| 1942 | War Production Board closes all gold mines to free up miners to work in strategic mineral mines. |
| 1971 | U.S. government abandons gold standard; gold prices rise dramatically. |
| 1975 | Ropes Gold Mine sold to Callahan Mining Corporation. |

Flotation cells used to recover valuable minerals at Callahan's Humboldt mill.
DAN FOUNTAIN

| | |
|---|---|
| 1980 | Curry shaft dewatered and Ropes mine evaluated. |
| 1983 | Sinking of new Ropes mine decline begins. |
| 1984 | Production of gold from the Ropes mine begins at the Humboldt mill. Vertical shaft excavated at Ropes. |
| 1987 | Old section of the mine caves into modern workings at Ropes Gold Mine. |
| 1989 | Callahan closes Ropes mine. |
| 2001 | Back Forty zinc, gold, silver and copper deposit discovered. |
| 2002 | Kennecott Minerals discovers nickel/copper deposit. |
| 2002-12 | Aquila Resources formed; explores the Back Forty deposit and forms plans to mine it. |
| 2010-11 | Aquila Resources diamond drills Peninsular gold property and finds gold up to $2,070 per ton. |
| 2010-13 | Kennecott develops Eagle Mine to produce nickel and copper, with platinum, palladium and gold as byproducts. |
| 2014 | Eagle Mine scheduled to go into production. |

# Chapter 1

 Early Gold & Silver Discoveries

French fur traders and Jesuit missionaries were the first known Europeans to visit the Lake Superior country, with the first, probably French explorer Étienne Brulé, arriving in 1622.

Exploring for minerals was not their aim, yet they couldn't help but notice the local Ojibwe people's use of copper implements and the masses of native copper they kept as talismans.

As early as 1636, an account published in Paris noted stories of "mines of copper" in the upper lakes in the New World. In 1738, Louis Denis, Sieur de la Ronde, explored the Ontonagon River area with two experienced German miners. They found not only copper, but indications of silver deposits along the Rivière Ste. Anne, now known as the Iron River. Although the French never developed the silver mines, the Iron River would become the center of a silver-mining boom more than a century later.

Interest in gold also came from early fur traders and entrepreneurs. In 1770, Alexander Henry, an Englishman who held the fur-trading concession for Lake Superior, entered into a partnership with fellow Englishmen Alexander Baxter and Henry Bostwick to prospect for minerals on the lake. Their first exploration was to search for the "island of yellow sands."

They had heard a local Native American legend that told of an island covered with heavy yellow sand and guarded by spirits. Hoping that this yellow sand was gold, Henry and his companions visited the island, now known as Caribou Island. In three days of searching, they found neither yellow sand nor vengeful spirits, much less any gold. Henry and company turned their attention to the masses of native copper to be found in the Ontonagon country.

A persistent legend claims that gold was first discovered in Michigan by Douglass Houghton, a young man whose short life would have a large impact on the region – with or without gold.

Houghton was a native of Troy, New York, and at the age of 20 graduated from the Rensselaer School, where he studied geology. He became an assistant professor at that prestigious scientific school shortly after graduation.

He was invited to Detroit, sponsored by the state Legislature, became a popular lecturer in the state and soon earned his license as a physician. Houghton served as surgeon and botanist on Henry Rowe Schoolcraft's expeditions to the headwaters of the Mississippi River and Lake Superior in 1831 and 1832, afterward reporting on the copper deposits of the Keweenaw Peninsula. Soon after Michigan achieved statehood in 1837, Houghton was appointed as the first state geologist by Governor Stevens T. Mason.

Over the next several years, Houghton surveyed and reported on aspects of the state's geology, including lakes,

A portrait of Douglass Houghton with his dog, Meeme, at Pictured Rocks, done shortly after his death by his brother-in-law, Alvah Bradish
MICHIGAN STATE ARCHIVES

rivers and salt wells. In 1840 he returned to the Upper Peninsula and studied the veins of native copper that had been discovered there. Houghton's report on these explorations provided the first scientific evidence of the mineral wealth of the Copper Country and spurred further exploration.

In 1844, Douglass Houghton convinced Congress to finance a joint geological and linear survey of Michigan. For the 1845 survey in the Upper Peninsula, Houghton worked with William Austin Burt, deputy land surveyor for the federal government. While Burt and his crew laid out the range and township lines and marked their corners, Houghton's men took detailed geological notes and samples and measured the land's elevation.

According to accounts published years later, Douglass Houghton discovered gold while camped near the present site of Negaunee in 1845.

The story has it that Houghton set off one afternoon on a solo excursion and returned with rock specimens carrying gold – some stories say enough free gold to fill an eagle's quill. Fearing that his men would desert to prospect for gold, he kept the find a secret. Houghton only revealed the discovery to his trusted associate, Samuel Worth Hill, the expert surveyor and mineral explorer whose penchant for spicy language has been immortalized in the euphemism "What the Sam Hill!"

Unfortunately, Dr. Houghton and two voyageurs drowned later that year when their Mackinaw boat capsized in a storm near Eagle River. The exact location of his find died with him. He was only 36 years old.

In June the following year, his younger brother, Jacob Houghton Jr., found a mineralized vein that held a small amount of gold at an undisclosed location in the Upper Peninsula. An assay to determine the ore composition and value done by Robbins and Hubbard of Detroit yielded 10¼ ounces of copper, 1¾ ounces of silver and 12 grains of gold from a 28-ounce specimen. The assay produced a gold bead, which Jacob Houghton wore for many years as the head of a stickpin.

Though Douglass Houghton had died, his reports of

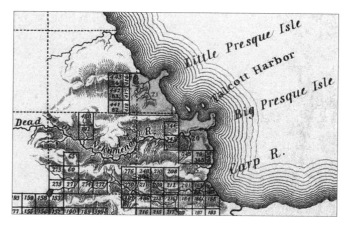

Marquette in 1844, showing Talcott Harbor, now known as Middle Bay

ALL IMAGES AUTHOR'S COLLECTION UNLESS INDICATED

the copper and silver wealth to be found in the Upper Peninsula continued to inspire interest and spurred a rush of prospectors into Michigan's Copper Country.

The influx was slow at first; the Upper Peninsula was still Indian land. The rush began in earnest with the ratification in 1843 of the Treaty of La Pointe, made with the Ojibwe people and which ceded to the United States the south shore of Lake Superior from the Chocolay River near Marquette to the St. Louis River at Duluth.

Under the mining laws of the day, government-owned mineral lands were under the jurisdiction of the War Department. The military was largely concerned with lead for ammunition and maintained a Mineral Agency based in Galena, Illinois, in the heart of the Illinois/Wisconsin/Iowa lead-mining district.

Shortly after the Treaty of La Pointe was ratified, the U.S. secretary of war instructed Walter Cunningham, the superintendent of mineral lands at Galena, to establish a new U.S. Mineral Agency at Copper Harbor near the tip of the Keweenaw Peninsula. The agency issued exploration permits to prospectors, allowing them to search for minerals. Once an explorer had located a promising site, he could lease up to 9 square miles of land under his permit.

Some of the early leases were issued to investors from the Albany, Troy and Watervliet area of New York, where Douglass Houghton had grown up and gone to school.

An artist's rendering of Camp Gray at Talcott Harbor.

On May 1, eight of these leaseholders formed the New York and Lake Superior Mining Company and assigned their leases to the company. Several of these leaseholders had connections to the War Department: Andrew Talcott was the brother of Lieutenant Colonel George Talcott of the Ordnance Office; Sebastian Visscher Talcott was George's son; Edward Learned Jr. was heir to a banking fortune and a distant cousin of the newly appointed secretary of war, William Learned Marcy. Other leaseholders included General Garret V. Denniston, James W. Glass, Radcliff Hudson, William J. Marlett and Benjamin E. Green.

The New York and Lake Superior Mining Company was organized by former congressman and artillery manufacturer Gouvenor Kemble of the West Point Arsenal in Cold Springs, New York, and he became its president. Other directors included *Albany Atlas* newspaper publisher H.H. Van Dyck, secretary; Albany banker Watts Sherman, treasurer; and trustees General Garret V. Denniston of Albany, New York, Edward Learned Jr. of Watervliet, New York, Andrew Talcott and William J. Marlett.

Andrew Talcott served as general agent for the company. Stockholders included others of national prominence, such as former U.S. President Martin Van Buren and Secretary of War William L. Marcy.

Lands held by the company included three tracts on the Montreal River along the border between Michigan and Wisconsin, three others on the Keweenaw Peninsula at

Eagle River, Agate Harbor and south of Copper Harbor, as well as two near the Dead River at present-day Marquette.

Over the summer of 1845, the New York and Lake Superior Mining Company's explorers, under geologist Dr. James Eights, examined the company's land for mineral deposits. The company's operations were based at their land at Agate Harbor, where a copper-bearing vein was soon discovered. Explorations on the Montreal River tracts also revealed several promising veins. Copper was also found on the Eagle River and Copper Harbor leases assigned to the company.

The Dead River tracts were the last to be explored, with operations beginning in August. One of these parcels, lease No. 20, was leased by Sebastian Visscher Talcott and encompassed Presque Isle, which was then known as Granite Point, and the mouth of the Dead River. At the north end of Presque Isle, the company's prospectors discovered numerous veins of galena, pyrite and chalcopyrite in the serpentine rocks. Since they were deposited under the same conditions as gold and silver, these minerals were often a clue to the presence of the precious metals.

In October 1845, a crew of miners was set to work to open a mine. The 15 men and two women, under the supervision of Superintendent Charles G. Learned (Edward Learned's older brother) and a Cornish mining captain, started construction on five log buildings, including a blacksmith shop and a storehouse.

Later in the month, the miners turned their attention to a narrow vein of galena, which, judging from its silvery luster, was believed to carry some amount of silver. Located south of the cove near Presque Isle's northern tip, the vein was explored by a shaft that was started a short distance back from the shore and about 20 feet above lake level. As the shaft was sunk, the galena, only about 2 inches wide at the surface, slowly gave way and was replaced by an equally narrow vein of "yellow sulphuret of copper" or chalcopyrite.

Finding minerals was only part of the battle. The New York and Lake Superior Mining Company found that securing reliable transportation for their miners and supplies on remote, unsettled Lake Superior was difficult. The company leaders decided to buy their own vessel. They purchased the schooner *Swallow* for $3,019.40 in June 1845, and equipped it with a pair of Ingersoll lifeboats to serve as lighters or barges.

    The schooner transported the company's people and supplies up the lake and carried the ore from their mines back down to Sault Ste. Marie. The *Swallow* also made money for the company by carrying freight and passengers for hire. To deliver supplies to the miners at Presque Isle, the schooner would pull into the sheltered bay formed by Presque Isle, Middle Island and Middle Island Point, which was then known as Talcott Harbor and now as Middle Bay. Since there was no dock, light cargo would be rowed ashore in the lifeboats, but when several head of beef cattle were brought to Presque Isle in the spring of 1846, they were simply pushed overboard and allowed to swim to shore.

    The shaft at Presque Isle eventually reached a depth of about 40 feet, at which point the galena had completely disappeared and had been replaced by chalcopyrite.

    The miners gave up and started a second shaft about 60 feet to the northwest. This shaft intersected the vein of galena about 10 feet down, but again only a narrow seam of mineral was found.

    One assay of the galena from the Presque Isle workings by Professor Jacob W. Bailey of West Point showed that the ore contained 70 percent lead, but no mention was made of silver. Work on the Presque Isle veins was apparently abandoned later in 1846.

    When the white men came to Michigan's Upper Peninsula in search of the frontier's treasures, the local Native Americans recounted legends of copper and silver and their connection with the supernatural. Sometimes those stories were thought to hold clues to the location of those riches. One such tradition told of a great cave lined with silver, which no white man had ever seen. Prospectors hearing this tale searched the area where the legendary cave was supposed to be hidden. Although the cave was never found, the explorers did find a large hill of basalt rising some 300 feet above its surroundings about 15 miles southwest of L'Anse.

Veins of "gray sulphuret of copper" (probably chalcocite, a gray or black sulfide) along a fault on the east side of the hill were promising enough that the National Mining Company of Pontiac, which was also exploring for copper in the Ontonagon district, took out a mining lease on the hill that would become known as Silver Mountain. The National Mining Company was organized by George W. Rogers, president; Henry C. Knight, secretary; banker James A. Weeks, treasurer; merchant Henry W. Lord, Henry B. Marsh and Jeremiah Clark, all of Pontiac, Michigan; and Detroit businessman Shadrach Gillett, who some years later would become Jacob Houghton's father-in-law.

Starting in 1845 or '46, National drove an adit (horizontal shaft) into the east side of the hill, following the vein. Although no records of the company's work survive, legend has it that the miners found native silver, which they melted down and cast into crude rings and crosses in molds carved from hard maple.

The adit was eventually driven 140 feet into the side of the mountain. At the end of the adit, the vein appeared to pinch out, so they sank a winze (a downward shaft) an unknown distance to see if the vein widened with depth.

The end of the National Mining Company's work at Silver Mountain came in 1847, and it was Lake Superior that changed the course.

The schooner *Merchant* left Sault Ste. Marie on June 12 that year, bound for Copper Country ports. Among the passengers were J.H. Woods and E. Gregory of Pontiac, Michigan, and L.C. Smith and George Howard of Norwalk, Ohio, all of whom were employed by the National Mining Company and were on their way to its location. When they were one day out from the Soo, a furious storm blew across Lake Superior. The heavily laden *Merchant*, passengers and crew were never seen again. The schooner apparently foundered in the storm, and the only trace of her ever found was a companionway door recovered on the Ontario north shore later that year. After that tragedy, operations at Silver Mountain apparently were never resumed.

A few years after National ceased work, "wreckers" came to Silver Mountain to salvage whatever they could from the abandoned mine. They hauled the anvil, blower and all the tools from the blacksmith shop about a mile to the Sturgeon River.

Orrin W. Robinson

Here they loaded their booty aboard a raft and drifted down the river toward Portage Lake. About 10 miles downstream from Silver Mountain, the Sturgeon River flows through Otter Lake. While crossing the lake, wind and waves caused the raft to go to pieces, sending its iron cargo to the bottom of the lake.

During the Marquette County silver excitement of 1864, Philo M. Everett, a pioneer mining man from Marquette, bought Silver Mountain, but never did anything to reopen the mine. No more was done with the mine until the 1880s when it was purchased by Orrin W. Robinson. Along with some Houghton and Hancock men, Robinson formed the Silver Mountain Mining Company in June 1888.

He assigned a one-half interest in the property to the partnership and was elected president of the company. E.L. Wright was secretary and treasurer, and Edward Ryan, J.C. Hodgson and William Condon were directors.

In early August, the Silver Mountain Mining Company sent William B. Davis and two miners to explore for minerals at the mountain. They spent two weeks at the mine and returned with a discouraging report. The Silver Mountain Mining Company was dissolved soon afterward.

Late in the 20th century, a study of Silver Mountain by the U.S. Geological Survey examined seven samples taken from the adit. Assays showed very low silver values; the highest assay yielded only 1/10th ounce of silver per ton.

One eastern businessman to try his luck with U.P. copper mining was Connecticut native Waterman Palmer, who was a successful merchant in Ohio and in Pittsburgh before coming to Michigan's Copper Country in the late 1840s. He served as the secretary and treasurer of the North American Mining Company in 1851. Three years later he was among the founders of the Central Mining Company, along with S.W. "Sam" Hill. He was also elected secretary and treasurer of that company, a post he held until 1859. Some time later, he relocated to the Marquette Iron Range to explore for iron ore, purchasing from the federal government large tracts of land south of the established iron mines. He found several deposits of iron ore and sold some of the land to the Cascade Iron Company and the Pittsburgh and Lake Superior Iron Company, which established the village of Palmer to provide the miners with houses and a company store.

In 1865 Palmer also found a vein of quartz that he suspected to be gold-bearing near the north shore of Palmer Lake on the SW ¼ of the NE ¼ of Section 22, T47N-R27W. An assay of one sample by Charles P. Williams of Philadelphia showed a value of $900.90 per ton in gold, while another assay by Julius Ropes of Ishpeming indicated $451.68 worth of gold per ton of ore. Neither Palmer nor the succeeding owner, the Pittsburgh and Lake Superior Iron Company, did any development work on the vein.

When the Empire iron mine open pit was developed in the 1960s, Palmer Lake and the adjacent gold-bearing quartz vein were buried under millions of tons of waste rock. Apparently other gold-bearing deposits exist within the mine pit, however. Over the years a few savvy workers at the Empire Mine have found pockets of heavy yellow mud deposited in low spots in the concentrator material flow – an unexpected gold mine within an iron mine!

## Chapter 2

 # The 1860s Silver Lead Rush

During two separate periods in the mid-1800s, the prospect of mining silver in the Upper Peninsula set off rushes, both generating plentiful speculation and formation of mining companies, but neither any really successful mining operations. The first of the rushes, involving the silver lead – the lead ore carrying silver – of Marquette County, was ignited by one man's discovery.

During the 1850s, Silas C. Smith, a pioneer prospector and mine owner, opened the first mine on the rich iron-ore deposit that came to be known as the Republic. Along with W.A. Pratt, he also established a whetstone quarry near the outlet of Teal Lake in Negaunee and set up a plant to cut and dress these stones in Marquette, near the mouth of a small stream that became known as Whetstone Brook. Smith, too, ended up being named for this business, earning the nickname of "Whetstone" Smith. A few months before his death, Smith claimed that he had also discovered mineral deposits containing gold and silver within the Marquette city limits in 1854 and 1855. Although Smith never developed these finds, later discoveries in the city supported his claims.

One of the men who worked for Smith on his prospecting ventures was a Native American whose name has been lost over time. This man had worked for some of the early iron-ore prospectors, including William and John Burt, and had gained a practical knowledge of minerals and ores. Several times he mentioned to Smith that he knew the location of a mineral deposit more valuable than copper or iron ore. Finally, in August 1863, he led Smith into a relatively unexplored area of Marquette County, north of what is now known as Silver Lake. Here, just below a waterfall on a small stream, he showed Silas Smith a wide vein of quartz richly charged with galena, a lead ore. From the appearance of the galena, Smith recognized that it held some percentage of silver.

Once Smith revealed "his" discovery, other prospectors began to search the surrounding area. Almost as fast as the explorers discovered new quartz veins, promoters founded new silver mining companies to develop them. The companies did little to test the veins, but rather offered stock for sale, printed prospectuses and published maps showing the various veins striking across the "Silver Range" for up to 10 miles, as straight as the ruler used to draw them.

One of the first companies formed was the Lake Superior Silver Lead Company, which was organized in New York to work Smith's original discovery. The founders were Alexander H. Sibley, president; Sylvester R. Comstock, treasurer; and Frederick B. Sibley, secretary.

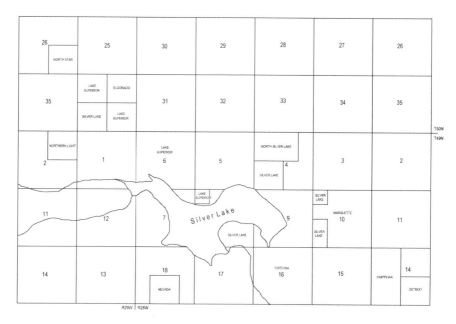

A map of covering the silver lead range of 1864

DAN FOUNTAIN

Trustees included these officers, as well as Nathaniel B. Palmer, James M. McLean, Edward Alburtis, David Smith, David Allerton and James Carson.

The partners were well positioned, both politically and financially. The Sibley brothers were the sons of Detroit's first mayor, Solomon Sibley, and their brother Henry Hastings Sibley had been the governor of Minnesota from 1858 to 1860. Comstock was the cashier of the Citizen's Bank in New York City, McLean was president of the Citizen's Fire Insurance Company, Smith was a manufacturer of lead products and Palmer was a veteran sea captain and owner of clipper ships. The company issued 200,000 shares of stock valued at $5 per share, including 20,000 non-assessable preferred shares to the founders. Treasurer Comstock bought Section 6, T49N-R28W, the site of Smith's discovery, along with another 390 acres along the presumed strike of the vein, for a total of 1,030 acres. He also purchased 135 acres at Big Bay and 78 acres on Huron Bay at Skanee to provide port facilities for the operations. Once the Lake Superior Silver Lead Company's organization was completed, Comstock transferred the property to the company. The company planned to start work that fall, but the heavy snowfall in the northern Marquette County highlands kept them from making much progress.

Alexander H. Sibley
ALL IMAGES AUTHOR'S COLLECTION UNLESS INDICATED

Among the first people to visit Silas Smith's discovery of silver lead near Silver Lake was John T. Martin of Houghton. Martin had been a miner and mine superintendent in England in his younger days. He visited the Smith vein in late September 1863, and searched for additional veins on adjacent lands. On the NE ¼ of Section 36, T50N-R29W, about a mile and a half from Smith's discovery, Martin discovered a 3- to 5-foot-wide vein of quartz, spar and chlorite running nearly north and south. When he broke loose some rock from the vein, he found galena, as well as copper and iron ores. Martin took samples of the galena and sent them to assayers in Hancock and Ann Arbor. He also purchased the land where the vein was located and returned to Houghton to organize a company to work the find.

On September 28, 1863, John Martin and seven other investors from the Houghton area formed the Eldorado Silver Mining Company.

The company issued 20,000 shares of stock with a par value of $25 per share. Martin received 6,000 shares, while the other founders – Columbus C. Douglass (Douglass Houghton's cousin and a Copper Country pioneer in his own right); Douglass's brother-in-law Ransom Shelden; John Williams of the Portage Lake Smelting Works; Captain William Harris, who was the mining captain of the Minesota copper mine; James H. Ralston; John Langdon and Albert Wanzer – received 2,000 shares each. Ransom Shelden was elected president and his business partner Albert Wanzer served as secretary and treasurer. Martin sold the NE ¼ of Section 36 to the company for $200.

The samples of galena taken from the Eldorado vein were assayed by Professor Silas H. Douglas, who had been Douglass Houghton's assistant in 1842 at the University of Michigan, and by John Williams of the Portage Lake smelter. The assay by Professor Douglas showed 3.26 percent silver in the galena, and Williams' analysis yielded 2.40 percent.

Other Copper Country businessmen were just as quick as the Eldorado partners to invest in the silver lead range. Douglass, Ralston, Wanzer and Captain Harris joined with C.C. Douglass's cousin, postmaster Edward F. Douglass, physician George Fuller and attorneys Thomas McEntee and James B. Ross to incorporate the Silver Lake Mining Company. The new company bought the quarter section of land adjoining the Eldorado and Lake Superior Silver Lead companies' lands on Section 36, T50N-R29W along with another 725 acres along the presumed strike of the silver lead veins. The company's 20,000 shares of capital stock, with a par value of $25 per share, were divided among the founders. A few of the stockholders made some of their shares available on the open market, letting it go at the bargain price of 25 to 50 cents per share to the eager buyers.

Edward Douglass, John Ralston, Thomas McEntee and George Fuller, along with furnaceman George Asmus of Houghton, formed another mining company a few weeks later. They founded the North Silver Lake Mining Company and purchased 200 acres of land just north of the Silver Lake Mining Company's property on Section 4, T49N-T28W.

Captain John Spaulding was a Great Lakes steamship captain who had commanded the propeller steamer *Northern Light* on the Lake Superior route for years. Spaulding's vessel freighted copper and ore down to lower lakes ports and merchandise up the lakes to the mining country, and carried passengers in both directions. This gave the captain plenty of opportunity to learn the latest news from the mining districts. Captain

The *North Star*

Spaulding was one of the early investors in land on the new silver lead range, buying up some 1,120 acres early in the fall of 1863. In November of that year, he and several partners incorporated the Northern Light Silver Mining Company. Investors included another steamship captain, Bemsley G. Sweet; Robert Hanna, the owner of the *Northern Light*; shipping agents H.J. Buckley and Bernard O'Grady; lawyer Andrew W. Buell; Benjamin T. Rogers, and Joseph M. Richards. The new company bought the NE ¼ of Section 2, T49N-R29W from Captain Spaulding.

The same group of investors also bought the SE ¼ of Section 26, T50N-R29W from Spaulding and formed the North Star Silver Mining Company, naming the company for Captain Sweet's longtime command, the sidewheel steamer *North Star*.

After the lakes shipping season closed, Captain Spaulding and Robert Hanna traveled to Philadelphia to attempt to interest investors in another mining company to explore for silver lead on their lands. They printed a prospectus that cited assays of ore from the veins on their land and other nearby properties. It also quoted reports from experienced mineral explorers and civil engineers, including George P. Cummings, Henry Merryweather and Homer Pennock.

Captain John Spaulding

Spaulding and Hanna were able to enter into a partnership with three Philadelphia merchants, Joseph Henszey, Edward H. Trotter and Jacob P. Jones. They incorporated a new company, the Chippewa Mining Company of Michigan. Along with Robert Hanna, each bought a quarter interest in the new company. The Chippewa company bought the W ½ of Section 14, T49N-R28W from Spaulding and Hanna and waited for the long northern Michigan winter to pass before starting operations in the spring.

Also in December 1863, another group of Philadelphia merchants founded the Marquette Mining Company of Michigan. Founders included Joseph L. Moss, William Morris Davis, Samuel M. Day and Charles C. Jackson.

The company bought 5,860 acres of land from the Lake Superior Land Company during the dead of winter, when deep snow blanketed the silver lead range. During the early months of 1864, the Marquette company bought another 4,880 acres of land, only to find out that they held more land than was legally allowed.

To split up their land holdings, the company formed four subsidiary companies, the Yellowstone, Alcona, Otsego and Monitor mining companies of Michigan.

The founders of the Chippewa and Yellowstone companies organized three other new silver-mining companies in March 1864: the Cherokee, Osage and Mohawk mining companies of Michigan. The same core group of investors in various combinations made up the stockholders, directors and officers of the new companies. Iron merchant Jacob P. Jones, hardware wholesaler Joseph Henszey, who was also the president of several Michigan copper mining companies, coal company executive James P. Jenks, coppersmith George R. Oat and metal merchants William H. and Charles W. Trotter, all of Philadelphia, controlled the new companies. The Cherokee company bought 1,000 acres of land in T51N-R28W, some 12 miles north of the Smith vein, and the Osage bought 760 acres in the same area, all during the winter months.

In the January 22, 1864, edition, Marquette's *Lake Superior News and Journal* noted the discovery of silver on the Salmon Trout River in northern Marquette County. The discovery was made by Norman E. Eddy, who had been guided to the site by a local Métis man. He was reported to have found a vein of silver crossing the river and running into high bluffs on either bank. The land where the vein was found belonged to Cullen C. Eddy, who soon sold the land to Charles M.W. Earle of Marquette. Two weeks after the discovery, Earle and investors from Detroit formed the Huron Mountain Silver Mining Company. The other founders were Anson H. Rood, Charles H. Wetmore, Henry K. Sanger and Henry P. Sanger, who was elected president of the company. Earle sold 200 acres along the presumed strike of the vein to the company for $25,000, probably paid in company stock. The parcel was in Section 29, T51N-R28W, some eight miles north of Smith's discovery.

The Marquette Silver Mining Company was organized by a diverse group of investors: Captain

Ransom Shelden

Bemsley Sweet and merchants John Holland and John E. Turner of Cleveland; Messrs. Buckley, Buell and O'Grady of the Northern Light and North Star Silver Mining Companies from Detroit; hotelkeeper Martin Foley of Copper Harbor; and storekeeper Benjamin T. Rogers of Rockland. The company bought 520 acres comprising most of Section 10, T49N-R28W in the rugged hills east of Silver Lake. Forty acres were purchased from Captain Spaulding while the rest had been jointly owned by Spaulding, Sweet, Holland, Foley and Rogers.

Samples from several quartz veins on the property were assayed but showed little galena and even less silver. One assay of pyrite in quartz by assayers DuBois and Williams of Philadelphia, however, showed a quantity of gold. The assayers reported that "the value of the ton of rock is above the average of that of Colorado." The company sank a shaft, test pits and trenches on at least three quartz veins across the section; one of them was described as a wide vein with pyrites carrying a small percentage of gold.

Following this revelation of gold in the silver lead range, other companies began to assay the yellow chalcopyrite and pyrite from their lands along with the silvery galena. There were reports of quartz veins carrying up to $700 per ton. Newly formed companies began to include "gold" in their names, such as the American Gold and Silver Lead Mining Company, incorporated by Ransom Shelden, William Harris and Albert Wanzer in March. Assay reports of rich gold veins were published in the press for a few months until it became apparent that there was nowhere near enough of the rich vein matter in any of these prospects to make gold mining a paying proposition.

Back in the Copper Country, it wasn't just the lawyers and wealthy professionals who were fueling the silver lead rush. Walter A. Northrup was a saloonkeeper in Hancock who was quick to buy up lands on the presumed silver lead range. During 1864 he bought more than 1,800 acres of what he hoped were valuable mineral lands, and formed at least six companies to explore for riches.

Along with John P.M. Butler of Houghton and William F. Hall of Keweenaw County, Northrup incorporated the Crystal Lake Silver Lead Mining and Smelting Company in January 1864. Butler served as secretary and treasurer, and mine agent Jonathan Cox was president. Northrup sold 320 acres in Sections 19 and 30, T50N-R29W to the company for $1,000. The south tip of the company's

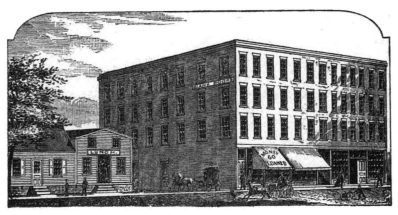

The Rotunda Building in Detroit, where many of the silver mine stockholders had law offices

land bordered on what is now called Bulldog Lake; perhaps this shallow, weedy lake was known as Crystal Lake in those days.

A few weeks later, Walter Northrup and nine other investors incorporated the Isabella Silver Lead Mining Company. Northrup and an associate, Michael Graveldinger, owned the majority of the stock. Northrup sold 600 acres of land to the company for $1,000. The land was in three parcels scattered across 10 miles of northern Marquette County. Northrup and Graveldinger were also the major stockholders in the Galena Silver Company, formed in May 1864. Six other Copper Country men comprised the rest of the founders. The company bought 320 acres of land from Northrup near his other companies in northern Marquette County.

Walter Northrup also organized the Jackson Silver Mining Company along with Abraham A. Jackson of Houghton and five other local investors. Northrup was happy to supply the new company with 400 acres of land along the east branch of the Salmon Trout River. The price for this parcel was a mere $400. Northrup retained 5,000 shares of the Jackson company stock, while the other stockholders owned 3,000 shares each.

Once the news of gold discoveries in the silver lead region became well known, Walter Northrup started to name his companies accordingly. Along with Michael Graveldinger, Chandler Kibbee and four other Houghton men, he formed the Cincinnati Gold and Silver Mining Company. Once again Northrup sold a promising parcel of land to the new company, receiving $400 for 320 acres. The lands spanned the Salmon Trout River just north of the Galena company's property in Sections 14 and 15, T51N-R28W. The Union Gold Company was incorporated by Walter Northrup, Jacob Shafer and five other Houghton men in May 1864. Shafer had bought the SE ¼ of Section 27, T52N-R29W from the government and transferred it to the company in exchange for 20 percent of the company's stock. The land was near the Little Huron River in Baraga County, some 17 miles from Northrup's other gold and silver companies.

George P. Cummings

The rush of prospectors and miners heading for the new silver lead range found rough going. There were no roads into this wild country, only the poorest of trails through the hills and swamps. Representatives of the silver-mining companies petitioned the Marquette County Board of Supervisors, citing the need for a road to the Silver Lake country. The county board agreed, and appropriated $6,000 from the road and bridge fund for the road's construction.

Over the winter of 1863-64, a route for the new Silver Lake Trail was surveyed, starting near the Bancroft Iron

21

Company's blast furnace at Forestville and crossing some 18 miles of rugged country to the Baraga County line. The route was blazed and used as a winter trail, with clearing and grading to be done once the snow melted in the spring.

George P. Cummings was a civil engineer and surveyor from Vermont who came to the Lake Superior country in 1857. In 1863, he was already engaged in mineral prospecting in northern Marquette County when he heard of Silas Smith's discovery, so he turned his attention to the search for silver-bearing veins. Late that fall, he discovered quartz veins carrying argentiferous galena about 15 miles east of the Smith vein. Investors were quick to buy land on the new east range and to organize mining companies.

Often these quickly formed companies attracted investors from outside the mining community. Thomas M. McEntee and James B. Ross practiced law in Detroit in the late 1850s. Their firm, Ross & McEntee, had an office in the Rotunda Building on

---

**C. M. W. EARLE,**
DEALER IN
**Boots, Shoes  Leather and Findings,**
*LAKE SUPERIOR, MICH.*
**MARQUETTE, Everett's Block.**    **ISHPEMING, Buckley's Block.**

---

Griswold Street in the heart of Detroit's financial district. After moving their law practice to Houghton in 1861, the partners handled cases for copper mining companies, and were well connected with mining interests when the silver lead boom began in 1863. Ross & McEntee invested in the Silver Lake, North Silver Lake and Nevada mining companies. When Cummings revealed his new discoveries, Thomas McEntee bought a quarter section of land on the new east range. Located on the NE ¼ of Section 20, T49N-R26W, the parcel held enough promise that McEntee decided to form a company to

search for silver lead. Rather than seeking investors in the Upper Peninsula, he offered the opportunity to his former co-workers and clients in Detroit. Along with 13 other investors, Thomas McEntee incorporated the First National Silver Mining Company. Investors included fellow lawyers from the Rotunda Building, a judge, a banker, an insurance agent, local government officials, merchants and officers of the board of trade.

Several other companies were organized to explore for riches on the east range. Lumberman Samuel S. Burt of Marquette heard of Cummings' discovery and located another vein nearby on the E ½ of the NW ¼ of Section 28, T49N-R26W. Burt bought the land and sent samples to be assayed in Detroit, revealing up to 4 pounds of silver per ton of ore. He recruited a group of Detroit businessmen, which included iron and shipping industry magnate Eber Ward, *Detroit Free Press* Publisher Duane Doty and brewer Nathan Williams to incorporate the Pioneer Silver and Lead Company. Samuel S. Burt also purchased the quarter section east of the Pioneer company's land, the NE ¼ of the section, and organized the Mystic Silver Mining Company. His partners in this company were brewers Nathan Williams, who served as president, and George C. Langdon, secretary.

B. Rush Livermore
LAKE SUPERIOR
MINING INSTITUTE

Farther northeast along the range were the lands of the Excelsior Silver Lead Company. This company of Marquette County men owned the SW ¼ of Section 22, where the vein had been traced by surveyor Leander C. Palmer. Founders of the company were iron and charcoal manufacturer Hiram A. Burt, his brother and business partner Alvin C. Burt, merchant Charles M.W. Earle, railroad man Jason McGregor and builder Jeremiah Van Iderstine. Marquette men also controlled the property east of the Excelsior's land. The Idaho Mining Company was founded by L.K. Dorrance, chief engineer of the Marquette and Ontonagon Railroad; Cornelius Donkersley, iron master of the Morgan Iron Company, tax collector Walter Finney, William O. Dwyer, and Samuel A. Forbes.

Early in the spring of 1864, the Marquette Mining Company of Michigan hired Homer Pennock to explore their lands for minerals.

Once the snow began to melt, it became apparent that much of the company's land was worthless for mining. Only 10 quarter sections were thought to hold any promise of minerals. The parcels were divided among the Marquette, Yellowstone, Alcona, Otsego and Monitor companies, each company receiving two quarter sections. Further explorations found neither silver nor lead on any of the lands, and the mining companies faded from the public's attention.

Captains Spaulding and Sweet of the Northern Light and North Star companies were among the first to start work on their prospects in the spring. They had equipment and crews of workers ready to head into the woods as soon as the weather would allow, eager to build their mining camps and to start mining the rich silver ore they were sure awaited them.

Over the summer, vague but flattering reports were heard from the prospectors. Construction of the new road to the silver lead range also had started in the spring. Work on clearing the route began in May and lasted most of the summer, with the road finally being declared passable for wagons in mid-September.

The Grant Mineral Land Company was founded in March 1864 in Philadelphia to invest in undeveloped land in Michigan's Upper Peninsula. The company built up their holdings by purchasing unexercised land rights that had been awarded to veterans, or their widows, of the War of 1812, the Black Hawk War and the Mexican American War.

While exploring in northern Marquette County in May 1864, George P. Cummings of Marquette found a mineralized vein of quartz 12 miles southeast of Smith's find, on Section 2, T48N-R27W, just north of the Dead River. The Grant company bought the land, comprising the N ½ and the N ½ of the S ½ of the section, and gave Cummings a quarter interest in it. Working with B. Rush Livermore, chief explorer for the Grant Mineral Land Company, Cummings continued to prospect the section. He was soon rewarded with the discovery of a vein of argentiferous galena with crystals ranging from the size of corn kernels to 2 inches square. An assay by John Williams of the Portage Lake Smelting Works in Hancock showed 144 ounces of silver per ton. The vein was also deemed rich enough in lead to mine profitably for the base metal alone.

Gathering support from investors from New York, Philadelphia and Marquette, Cummings, along with druggist Henry H. Stafford, Andrew G. Clark and Sydney E. Church of Marquette, organized the Holyoke Mining Company on August 22, 1864. Henry R. Mather was elected president, and Church was secretary and treasurer. Directors included George Trotter of Philadelphia, Charles P. Martin of New York, and Cummings, Livermore, Samuel P. Ely and Peter White, all of Marquette. The new company set up an office in Church and Mather's wholesale liquor store in Marquette and issued 20,000 shares of stock at $25 each. The Holyoke company also purchased the Grant Mineral Land Company's land on Section 2 and bought the remaining 160 acres of the section from a war widow.

Little was heard from the Silver Lake Mining Company over the spring and summer of 1864. Apparently the company was waiting for further discoveries on the range to guide them in their explorations.

Finally in September 1864, the directors called in a 6 cents-per-share assessment to finance the hiring of a professional mineral prospector. Mr. F. Fohr of Houghton was employed to explore the company's lands. Fohr found two quartz veins that carried chalcopyrite and a small amount of galena on Section 4, T49N-R28W. Assays showed a good percentage of silver in the ore, but the specimens of galena were referred to as "the size of a pea" – hardly the hoped-for bonanza. Investors' confidence in the Silver Lake Mining Company must have been waning, as some 600 shares of stock were declared delinquent for non-payment of the September assessment. The delinquent shares were sold at public auction, bringing in from 33 to 78 cents per share. This was the last mention of the Silver Lake Mining Company in the press.

Fohr also evaluated the lands of the North Silver Lake Mining Company. The company published assay reports showing silver and gold values from its veins, but, like its predecessor, faded from public view and was soon forgotten.

A horse whim used to raise buckets of ore and rock from the mines.

Summer of 1864 found the Lake Superior Silver Lead Company at work on the Smith vein under Captain Thomas Low. The miners started an adit into the vein on the north side of the stream bed. This was driven some 50 feet into the vein, at which point a winze was sunk about 15 feet. From the bottom of the winze, a drift followed the vein another 125 feet. The vein was found to strike 18 degrees west of due north and was 15 feet wide at its widest point. On the south side of the stream, opposite the adit, an inclined shaft was sunk to an unknown depth in the vein. By trenching and test pitting, the vein was traced several hundred feet north of the stream, where another shaft was sunk. The final depth of this shaft is unknown, but it must have been considerable, since the remains of a whim (a horse-powered hoist) and the connecting rods of a pump could still be found at the site late in the 20th century.

Although the Lake Superior Silver Lead Company's workings never progressed to the point of installing a stamp mill, some attempt was made to separate the galena from the quartz later in 1864. Ore from the shafts and adits was crushed and concentrated on washing floors, crude gravity separators patterned after those used in the Keweenaw copper mines and the tin mines of Cornwall, but there are no records of the mine ever shipping any ore or metal.

There had been little mention of the silver-mining companies in the newspapers over the summer of 1864, but as the season wound down, disappointing news began to surface. In September a legal notice was published that a large amount of the stock of the First National Silver Mining Company was to be sold for non-payment of an earlier assessment. Explorers on the Idaho company's property reported poor success. The Marquette Silver Mining Company called an assessment on its stock in September, but in late October it was noted in a local newspaper that work there had been suspended. The Chippewa Mining Company had opened several trenches exploring quartz veins and had sunk a shaft on the most promising one, but it too ceased operations in October. Work at the Lake Superior Silver Lead Company's mine was also suspended for the winter. The Eldorado company had called a 5-cents-per-share assessment on its stock in September, but little was paid in. The company tried to sell the delinquent stock in December, but was never heard from again.

The Crystal Lake Silver Lead Mining and Smelting Company called a 2-cents-per-share assessment on its

Homer Pennock

stock in mid-November 1864 and scheduled their annual meeting for February 1865, but there was no further mention of the company in the local press. No mention was made of Walter Northrup's other companies, the Isabella, Jackson, Galena, Cincinnati, and Union companies, so it is unknown if any mining was done or if any gold or silver was ever found on these properties.

Walter Northrup later moved on from the Michigan's Copper Country, but in 1872, he and Homer Pennock were back in the news as the principal players in what became known as the Otter Head Tin Swindle. Northrup and Pennock had supposedly discovered rich veins of tin ore on the wild Canadian north shore of Lake Superior and had bought up thousands of acres of land in the vicinity. Northrup obtained the backing of a group of investors from Detroit who formed the Otter Head Tin Lands Pool and bought Northrup's and Pennock's land holdings, paying them in cash and stock in the pool. The investors engaged Captain William Harris of the Minesota copper mine to inspect the veins.

Homer Pennock guided Captain Harris to the tin lands in November 1872, just before the close of

**Holyoke shaft**
LIBRARY OF CONGRESS

Miners' camp at Holyoke
SUPERIOR VIEW

navigation. Harris had only a short time to inspect the veins and had to dig through snow to secure samples. Upon bringing ore specimens back to Michigan and testing them, Harris found that they did indeed carry valuable amounts of tin. With this proof, Captain Harris endorsed the find as genuine. Harris was a highly experienced, well-respected mining man from Cornwall, and his endorsement of the find was all it took to create a demand for Otter Head stock in the eastern markets. Northrup and Pennock quietly sold all of their stock.

In the spring of 1873, once the snow was off the ground, mineral experts from Ontario's Fort William and from Silver Islet inspected the tin-bearing veins and discovered that the veins had been "salted" – the original matrix removed and replaced by artificial stone containing a generous amount of genuine high-grade tin ore. Harris, too, revisited the site and found the chemicals used to salt the vein hidden nearby. By this time, Pennock and Northrup were long gone from the mining country. Homer Pennock eventually went to Alaska during the gold rush of 1896, where he founded the town of Homer. Walter A. Northrup, too, followed the precious metals. The 1880 census shows him living in

Leadville, Colorado, during the height of the silver boom there. He died in 1895, noted with a vague reference in the Northrup family genealogy that he "was killed in the West. Was 'salting mines.'"

Despite the promise of its mineral veins and the strength of its financial backing, the Lake Superior Silver Lead Company's operations were never resumed after the mine closed down for the winter of 1864-65. After the closing of the Lake Superior mine, Alexander Sibley continued his involvement in mining, investing in the phenomenally rich Silver Islet silver mine on the Ontario north shore of Lake Superior, eventually becoming president of the company. Another major investor in the Silver Mining Company of Silver Islet was Edward Learned Jr. of the short-lived New York and Lake Superior Mining Company, who owned a quarter interest in the company and succeeded Sibley as president.

Hand-powered blower
LIBRARY OF CONGRESS

While many of the early silver lead-mining companies were failing in the fall of 1864, operations at the Holyoke mine were just beginning. Mining began in September 1864 under the supervision of Captain Henry Slockett. An assessment of 50 cents per share was levied to finance setting up the camp and sinking the shaft. Sinking commenced at what became known as the No. 2 shaft, situated where two of the veins on the property intersected. A crew of eight woodsmen, two carpenters and a teamster with a yoke of oxen built a large log mess hall, three smaller cabins for miners' living quarters, a blacksmith shop and a stable. They also cleared a wagon road from the Silver Lake Trail to the mine, which became known as the Holyoke Trail.

The silver lead ore seemed to be plentiful at the Holyoke mine. During the first three months of mining, five barrels of clean ore were shipped, each barrel weighing more than 600 pounds. By the end of January, 1865, the No. 2 shaft had reached a depth of about 40 feet, but the vein had nearly pinched out. Operations were shifted to the No. 1 and No. 3 shafts on a smaller vein some distance to the south. Although this southern vein was richly mineralized, it was too narrow to be worked at a profit. In hopes of finding the vein to be wider at depth, the company decided to drive an adit from the level of the camp below the bluff. Driving the adit, as well as further sinking in the shafts and the construction of a sawmill, was financed by a $2 per share assessment called in January 1865. The adit was started into the side of the bluff perpendicular to the south vein and struck it about 150 feet in. At this depth, more than 100 below the top of the ridge, the vein was still too narrow to pay. The adit was extended toward the northern "main" vein, which it was expected to intersect at about 350 feet.

As the adit grew deeper, ventilation became a problem. At first a hand-cranked blower was used to force fresh air to the end of the adit through 4-inch-square wooden ducts. The manual blower soon proved inadequate, however, so Jake Humbert, the mine's carpenter, built a 16-foot-diameter water wheel to power a larger ventilator. Water was taken from a beaver pond on a small stream north of the No. 2 shaft to turn the wheel, which also powered a roller crusher for pulverizing the ore.

The adit passed the 350-foot mark in July 1867 without a trace of the "main vein" being found. The miners returned to the No. 2 shaft and started a drift, which followed the vein more than 100 feet to the west. Careful measurements of the vein were taken, which revealed that the adit should intersect the main vein at least another 100 feet farther into the ridge. Work was continued on the adit, and another shaft, the No. 4, was started 500 feet from No. 2 shaft, directly above the point where the adit was predicted to intersect the vein. The vein here was 7 feet wide and richer than anywhere else on the surface. Work in the adit was suspended during the winter, the bitter temperatures causing the water wheel and ventilator to freeze up. The next spring, work in the adit resumed, driving it to its final depth of 637 feet without ever striking the elusive main silver vein.

In the course of sinking the shafts, drifts and adit, more than 100 tons of ore had been taken out. Captain Tom Allen, the veteran Cornish tin miner who had succeeded Captain Slockett, felt that he could extract the silver. The ore was crushed in the water-powered roll crusher, but no attempt was made to separate the silver-rich galena from the quartz. The crushed vein rock – galena and quartz alike – was put into an adobe smelting furnace and heated with a charcoal fire. Three attempts proved unsuccessful, however. In one attempt, the half-molten ore "chilled" (solidified) in the furnace, and part of the furnace had to be torn down to free the hardened mass. The other attempts supposedly "burned up" the silver and lead.

John H. Forster

At the Holyoke silver mine, work was suspended in October 1868, when a lack of operating capital forced its closure. Stockholders had tired of repeated assessments on their stock and the lack of silver production from the mine. Unpaid creditors took the company to court in February 1869, and the assets of the Holyoke Mining Company were sold at a sheriff's auction in April. The Wetmore brothers of Marquette were the successful bidders, but never did anything to reopen the mine. Several attempts were made by other parties to reopen the mine, notably in 1885 and 1899, but no more production was recorded. The buildings provided shelter for trappers and fishermen for a few years, and the adit was a lure for the adventurous until the 1960s when it was capped and tons of rock bulldozed over the entrance.

A clue to the ultimate disposition of the silver-bearing ore from the Holyoke mine was discovered in the 1980s. While walking near the old Greenwood blast furnace west of Ishpeming, a Negaunee man found a rough disk of gray metal several inches across protruding from a washed-out earthen bank. Tests by a chemist proved the metal to be nearly pure silver. This "button," which weighed more than 3 pounds, still showed bits of

## PROSPECTUS

FOR THE FORMATION OF THE

### 𝕷ake 𝕾uperior 𝕾ilver-𝕷ead 𝕮ompany,

SECTIONS 6 & 36, T. 49 & 50, R. 28 & 29.

*Marquette County, Lake Superior, Michigan.*

---

charcoal on its upper surface, as well as the marks of the crude sand mold in which it was cast. The Greenwood Furnace was a charcoal fueled blast furnace built in 1865 by the Marquette and Ontonagon Railroad. This pig iron-producing furnace was sold in 1868 to the Michigan Iron Company. Andrew G. Clark, the secretary and treasurer of the Michigan Iron Company, was also one of the founders and a major stockholder in the Holyoke Mining Company. It is certainly possible that, after Captain Allen's unsuccessful attempts to refine the Holyoke ore and the failure of the company, Clark decided to try to recoup part of his investment by recovering the silver from the remaining ore. The argentiferous galena would have been smelted to yield silver-bearing lead, which could have been easily refined to extract the silver.

The Holyoke silver mine also gave rise to one of the classic "lost mine" stories of the Upper Peninsula mining country.

One of the miners at the Holyoke was Mose Williams, who later became mining captain at the Ropes Gold Mine. While grading a wagon road near the Holyoke mine in the 1860s, Williams and two companions found a rich vein of galena just below the surface. The men decided to cover the vein and keep it a secret until they could lease or buy the land where it lay. After the mine closed, they tried to get control of the land, but it could not be secured. Years later, after the deaths of his partners, Mose Williams returned to the Dead River area to try to find the lost vein. He searched for a week, but the appearance of the land had changed so much that he could not find the vein of galena they had hidden some 20 years before.

In all, more than 50 silver lead-mining companies were organized in 1863 and 1864. Some were named for the hometowns of their founders, such as the Agawam Silver Mining Company, backed by investors from Agawam, Massachusetts; the Holyoke Mining Company, for neighboring Holyoke, Massachusetts; the Keweenaw Silver Mining Company, the L'Anse Silver Mining Company, the Portage Lake Mining Company,

the Marquette Silver Mining Company, the Detroit and Marquette Silver Lead Company, the Detroit City Silver Mining Company (whose incorporators included Detroit Mayor William C. Duncan), and the New York and Michigan Silver Lead Company.

Others were given names that seemed to conjure up visions of fabulous wealth – the Eldorado Silver Mining Company, the Fortuna Mining Company, and the Golconda Mining Company. Native American tribes, both local and distant, were popular namesakes for mining companies, including the Chippewa, Cherokee, Osage, Mohawk and Sioux mining companies. The Shuneaw Silver Mining Company's name was a phonetic spelling of the Ojibwe word for silver.

Surveyor, engineer and State Senator John H. Forster of Houghton bought several sections of land in southeastern Houghton County (now Baraga County) in the early days of the silver boom, and he organized companies to explore the lands for silver. He showed his appreciation for literature by naming the companies the Valjean, Fantine, Cosette and Marius Mining Companies, after the protagonists in Victor Hugo's classic novel *Les Misérables*.

The Holyoke Mining Company was the last of the 1860s silver lead-mining companies to close down their mining operations, but several companies stayed in business, though not mining, for some years longer.

The Marquette Mining Company of Michigan continued its corporate existence until at least 1883, calling in assessments on its stock in 1865, '68, '69, '74 and '83, and trying to sell delinquent shares and its land holdings. The Osage Mining Company continued to hold annual meetings and file annual reports with the state government through at least 1876, stating "no product to report."

None of the companies created in the 1860s silver lead rush would succeed as mining operations in the Upper Peninsula, but that did not deter another rush for silver that was soon to follow.

## Chapter 3

 The Ontonagon County Silver Rush

Ontonagon County in the 1860s was wild, remote country even by 19th-century Upper Peninsula standards. A few copper mines had been opened, some as early as 1847, and prospectors roamed the ridges and valleys looking for the next copper bonanza. One such prospector was Austin Corser, a native of Vermont, who had come to the Lake Superior country seeking his fortune in the 1840s. In 1847 and 1848 he was the first – and only – postmaster of the short-lived settlement of West Presqu'ile at the mouth of the Presque Isle River. He and his brother settled in Ashland, Wisconsin, in the 1850s. Leaving his wife and infant son in Ashland, Austin Corser prospected in the back country of Ontonagon County. Sometime around 1855, he discovered a mineralized vein in the bed of the Little Iron River (also known as the Pe-wa-bek, Ojibwe for "iron"). The vein wasn't iron ore, however, but proved to be silver-bearing quartzite.

Corser was hardly the first to find silver in the Ontonagon area. Louis Denis de la Ronde had sent a pair of German mining experts, John and Christopher Forster, to examine the Ontonagon country for minerals in 1738. They returned from the Rivière Ste. Anne, as the French knew the Iron River, with reports of copper deposits and of a rock formation favorable for silver. Other prospectors had found scattered pieces of silver ore, but nobody before Austin Corser had ever found the vein in place.

Once he had determined the extent of the vein, Corser proceeded to the government land office to buy the parcel. Federal land could be purchased from the government for $1.25 per acre under the Land Act of 1820. Another government policy, however, granted federal lands to railroad companies to finance the construction of their lines. This was the fate of the section where Corser's silver vein lay. It was part of the 142,430 acres that had been granted to the newly incorporated Ontonagon and State Line Railroad, which proposed to build a rail line from Ontonagon to the Wisconsin border. Although the railroad company wouldn't gain ownership of the land until the railroad was actually built, the land was not available for sale. Corser decided to wait out the railroad, gambling that the line would never be built. He built a cabin near the vein and moved his family from Ashland onto the land. Here the Corsers lived, the family increasing to six with the births of three more children. Austin Corser continued to prospect for minerals as well as farming the land while waiting for the railroad's

Ontonagon Silver Range
DAN FOUNTAIN

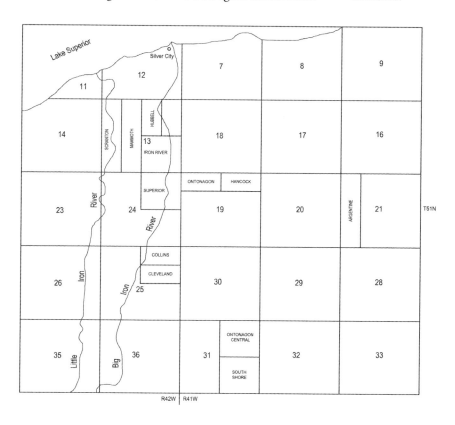

grant to expire. The railroad never did get built, and in 1872 the land reverted to federal ownership and became eligible for homesteading. Corser filed a preemption on the land and at last became the owner of the silver vein. He secured a patent on the W ½ of the W ½ of Section 13, T51N-R42W, where he had first discovered the vein. He had also traced the vein onto the E ½ of the SE ¼ of the section, and had found the silver formation on the SE ¼ of Section 18, T51N-R41W, both of which he and his wife, Sarah, bought in her name.

**T. MEADS,**
—DEALER IN—
**SPECIMENS, CURIOSITIES,**
Jewelry, Notions and General Merchandise.
Large assortment of Lake Superior Birds and Animals on Exhibition.
Papers and Periodicals for sale.     Marquette, L. S., Mich.

ALL IMAGES AUTHOR'S COLLECTION, UNLESS INDICATED

For all the years that Corser had homesteaded on the land, the silver vein had been a loosely kept secret. He had let a few close associates in on the news of the silver discovery. One who saw the potential of the find was Daniel Beaser, a retired sea captain who had settled in Ontonagon in its pioneer days. During the 1860s, Captain Beaser quietly bought up all the land around the mouths of the Iron and Little Iron rivers. He also bought land along the presumed course of the silver vein.

While Corser was the explorer, Beaser was the promoter. Along with Alfred Meads, the publisher of the *Ontonagon Miner*, Beaser was tireless in bringing the Ontonagon area's potential for silver mining to the attention of investors. They sent specimens of the silver ore to Meads' brother, Thomas Meads of Marquette, who ran a museum and store that dealt in mineral specimens. Thomas Meads had one of the first specimens of Iron River silver ore assayed by Professor F.B. Jenney of the Michigan Geological Survey at his lab in Marquette. The assay showed $206.40 worth of silver per ton of ore. Chemist Julius Ropes of Ishpeming assayed other specimens, one of which yielded $297.86, another $116.69 and a third $57.38 in silver per ton.

Throughout the fall of 1872, mineral experts, mining men and potential investors visited the surface exposure of the silver vein. In October, Captain William Frue of the fabulously rich Silver Islet mine in Ontario and Captain William Harris of the Minesota copper mine escorted a group of English investors to the vein and pronounced it a valuable find. Captain Beaser traveled to New York and showed specimens of the rich ore to eastern investors, backed up by a recent assay that had yielded $1,792 per ton.

By December, much of the land near Corser's find had been bought up, largely by Marquette investors. Exploring parties set out to locate the silver veins, but by this time the snow was too deep to follow the geological features. Despite the slight amount of exploring that had been done, investors had enough confidence to form mining companies and plan their mines. A group of Marquette men organized the Ontonagon Silver Mining Company in February 1873. Marquette alderman and postmaster Terrence Moore served as president, Deputy Collector of Internal Revenue Walter Finney was vice president, photographer Brainard F. Childs was secretary and Richard P. Traverse was treasurer. Moore and Traverse had purchased the N ½ of the NW ¼ of Section 19, T51N-R41W, a mile east of Austin Corser's silver discovery in the fall of 1872, and had been able to trace the vein onto their land before the winter snows. The Ontonagon company put 1,000 shares of stock on the market to raise capital and had no problem selling them all at $5 per share. Within a week, eager investors were offering up to $10 per share, but few stockholders were selling.

Although the snow still lay four feet deep in the woods, the Ontonagon Silver Mining Company set crews to work in mid-March. The silver range lay some 12 roadless miles from Ontonagon, so the men travelled by snowshoe, packing in everything they needed on their backs and on

BLAKE'S STONE AND ORE BREAKER.

dogsleds. At first the crew lived in a camp of crude brush shelters until they cleared the land and built log cabins for more permanent accommodations. After clearing away the snow, they dug trenches and test pits to expose the vein. They found the silver-bearing quartzite vein to be 17 inches thick, lying between layers of slate above and sandstone below and dipping at a shallow angle to the northeast. The silver was found as tiny, sometimes microscopic, flakes of native silver in the vein rock.

The Ontonagon Silver Mining Company continued to sink their exploration pit on the vein, timbering it for support and turning it into a proper shaft. However, the shaft was only about 300 feet from the property line, and it soon became obvious that the shallowly dipping mineral vein would pass onto the neighboring property before it was deep enough to develop into a producing mine. A second shaft was begun 200 feet to the southeast along the vein. The Ontonagon company hired Thomas Hooper, a Cornish mining captain, to supervise the shaft sinking. Captain Hooper set two shifts of miners to work, sinking day and night. As they mined deeper into the vein, it was discovered that some of the slate hanging wall – the layer of rock immediately above the silver vein – also carried up to 10 percent silver, bringing the total width of pay rock to 24 inches. At a depth of 115 feet, the miners started drifting east and west along the vein to open up pay ground for production. The vein here was found to be wider and richer than the surface showings.

Captain Beaser was also quick to start work on his

prospect. Beaser had purchased the NE ¼ of Section 24, T51N-R42W in 1865, shortly after learning of Austin Corser's discovery. The silver vein was exposed in several places on his land. Beaser started sinking on the vein where it outcropped on the east bank of the Iron River. Here the vein dipped at such a shallow angle that the mine opening was more of an adit than a shaft. Some of the same investors who made up the Ontonagon Silver Mining Company bought Beaser's prospect and organized the Superior Silver Mining Company to develop it. Terrence Moore was president, iron mining man and Marquette Mayor Samuel P. Ely was treasurer and Captain Beaser was appointed superintendent.

By early July, the Superior Mine had mined enough silver ore that the management decided to ship a sample to a refinery for a test run. The 1,200 pounds of ore that had been raised during the driving of the adit was packed down the trail to the mouth of the Iron River, where it was barreled and shipped to Ontonagon aboard the small sloop *Phantom*. There it was loaded aboard the steamer *Keweenaw* and delivered to the Wyandotte Silver Smelting and Refining Works near Detroit. The Wyandotte Works had been built just two years earlier to process the silver ore from Silver Islet on Lake Superior's Ontario north shore. Assays of the first shipment of ore showed from $11 to $1,716 per ton. The Superior company made several more shipments of ore to Wyandotte over the summer of 1873 for a total of about 2 tons, which yielded $632.71 per ton in the refinery. Plans were announced for a second shaft to be started 300 feet southeast of the adit.

Marquette dry goods merchant Charles G. Collins was an early investor in the silver range, buying 400 acres in Section 25, T51N-R42W a mile south of the Superior mine. In June 1873, Collins and several other

> HIRAM A. BURT,
> 
> **Land and Real Estate**
> 
> DEALER.
> 
> **Residence Property**
> 
> Upon nearly every Street, and in every part of the
> 
> CITY OF MARQUETTE,
> 
> ALSO,
> 
> **Choice Business Locations**
> 
> AND MANUFACTURING SITES
> 
> For Sale upon the most Liberal Terms to Purchasers for improvement.
> 
> Office Phoenix Block, Main Street,
> 
> **MARQUETTE.**

Marquette businessmen incorporated the Ontonagon and Lake Superior Silver Mining Company. Collins took an active role in the company as president, while Peter C. Beanston served as vice president, Edward B. Gay was secretary and Richard P. Traverse was treasurer. The company bought the N ½ of the NE ¼ of Section 25, T51N-R42W from Collins. They were able to raise working capital by selling 30,000 shares of stock to eager investors for $2 per share. By August the Collins company, as the Ontonagon and Lake Superior Silver Mining Company was commonly known, had a crew of miners sinking a shaft on the silver vein, working under another Cornishman, Captain Moyle.

Edward C. Anthony

Later in 1873, two new silver mining companies were organized by Marquette physician Samuel D. Humphrey. Humphrey was something of a Renaissance man. He had been the publisher of *Humphrey's Photographic Journal* and was a recognized authority on the art and practical science of daguerreotype photography. At the age of 40, he took up the practice of medicine, and he served a term in the New York State Assembly before moving to Michigan in 1871. While living in Marquette, he served as the county superintendent of the poor as well as being president of the Michigan State Medical Convention for three years.

With his wealth and his acquaintances in the professional community in Marquette, Dr. Humphrey was in a good position to capitalize on the Iron River silver discoveries. He bought 1,600 acres of land in Ontonagon County near the known silver veins in 1873. Much of it proved to be worthless for mining, but two quarter sections in Section 31, T51N-R41W showed enough promise that Humphrey and several associates incorporated mining companies to explore them.

The Ontonagon Central Silver Mining Company was organized to explore the NE ¼ of Section 31, two miles southeast of the Collins mine. Dr. Humphrey was president, and banker C.M. Everett, lawyer E.J. Mapes, J.M. Gordon, and J.M. Johnson served as directors. Humphrey and Mapes also joined with Ishpeming photographer A.G. Emery, Negaunee attorney M.H. Crocker and harnessmaker E.C. Anthony to form the South Shore Silver Mining Company. The South Shore company bought the SE ¼ of Section 31 from Humphrey. The two companies hired an experienced mining man, Captain James Williams, to direct the exploration of their lands.

Prosper Pennock of Ontonagon bought the E ½ of the W ½ of Section 13 adjacent to Austin Corser's original silver discovery early in 1873. He had traced Corser's vein onto the property and was able to interest Terrence Moore of the Ontonagon and Superior companies in buying the land. Moore and his associates organized the Mammoth Silver Mining Company to prospect and develop the property. Moore took his usual position as president, while his assistant postmaster, Washington J. Beardsley, served as secretary and treasurer. Prosper Pennock was hired as superintendent, as well as being vice president. Pennock's crew sank a shallow shaft and produced rich specimens of vein rock.

The Iron River Silver Mining Company was organized in Marquette in May to develop Austin Corser's silver find on the Iron River. Founding investors included George Northrop, David Murray, Byron P. Robbins, and Hiram A. Burt of Marquette, and Charles C. Briggs of Pittsburgh. The company bought the SE ¼ of Section 13, T50N-R42W from Austin and Sarah Corser. Perhaps it was no coincidence that this was the location where Hiram Burt's brother John claimed to have found silver ore more than 25 years earlier.

Alfred P. Swineford

In October 1873, a buyer was finally found for Austin Corser's original find of silver ore on the Little Iron River. The Scranton Silver Mining Company was formed by Michigan and Pennsylvania parties, who bought the W ½ of the W ½ of Section 13, T51N-R42W from

Charles Hebard

Corser. John Spaulding, the Great Lakes steamer captain who had invested heavily in the Marquette County silver lead range some 10 years earlier, was one of the founders and served as president. Lawyer Charles Van Fleet of Scranton, Pennsylvania, was secretary, and coal mine developer Nathaniel Halstead served as treasurer. Other Pennsylvania founders included Edward Dolf, a coal mining and lumber baron also from Scranton, and Charles Hebard, a wealthy lumberman who later became one of the biggest timber producers in the Upper Peninsula. Henry C. Thurber, an iron mining agent, and Alfred P. Swineford, publisher of the *Marquette Mining Journal*, were the Michigan partners in the company. Austin Corser received 2,000 shares of the company's stock.

The Scranton Silver Mining Company hired Captain Hooper to head up their exploration and mining efforts. Hooper set crews to work clearing the land and erecting accommodations for the miners. They planned to sink two shafts, an inclined shaft following the vein to prove up the mine and a vertical shaft 275 feet away to serve as a production shaft once greater depths were reached. The company also planned to build a mill to extract the precious metal from the ore.

By the fall of 1873, the Ontonagon Silver Mining Company had cleared and graded a wagon road from the mouth of the Iron River to their mine. Several other companies were exploring along the silver range. The Argentine Silver Mining Company of Marquette prospected for the silver vein on their property, the W ½ of the W ½ of Section 21 T51N-R41W. The Hancock Silver Mining Company explored their 80-acre parcel adjacent to the Ontonagon silver mine on the east. Although the silver vein did not reach the surface on their land, they planned to sink a shaft to intercept the vein at depth. Explorers working for copper mine owner Jay Hubbell on the W ½ of the NE ¼ of Section 13 east of the Mammoth property found the vein and started sinking a shaft near the Iron

River. Despite a promising showing of silver at Corser's find along the river in the SE ¼ of the section, the Iron River Silver Mining Company never did find a vein either wide enough or rich enough to develop.

Just as the dreams of silver mines in the Iron River District seemed poised to come true, global economic forces took a serious toll on the silver companies' ability to raise the capital they would need to develop producing mines. Changes in monetary policies in Germany and the United States had eventually triggered the failure of a major American banking firm in September 1873, marking the beginning of the Panic of 1873 and a depression that would last the rest of the decade. Banks and railroads across the country failed and credit dried up. Most of the silver mines and prospects in the Iron River district shut down operations for the winter.

The spring of 1874 found little activity on the silver range. Work at the Collins mine had been suspended the previous fall. The Superior mine likewise had been inactive over the winter, and the company was trying to make arrangements to pay off its debts. At the Ontonagon silver mine, only a few men were still employed, keeping the mine pumped out.

Only the Scranton Silver Mining Company was doing any underground work. Their shaft had reached the first level, where the miners were drifting east and west along the vein. Both the Scranton and the Ontonagon companies were considering installing processing plants to separate the fine-grained silver from the ore, despite the opinion expressed by some experts that it would be more cost-effective to ship the ore to New York to extract the silver. In July, officials from the Scranton Silver Mining Company visited the silver range. They expressed their confidence in the mine and came to an agreement with the Ontonagon Silver Mining Company to jointly finance a reducing and refining plant to process silver ore from both mines.

Jay Hubbell
LIBRARY OF CONGRESS

## JAMES PICKANDS & CO.

DEALERS IN

### Heavy Hardware,

**MINING AND RAILWAY SUPPLIES,**

Bar, Band and Sheet Iron,

NAIL, SPIKE. NUTS AND WASHERS,

### English and American Steel

Gas Pipe and Steam Fittings.

Steam Hose,     Hydrant Hose.
Suction Hose,     Miners' Handles,

**Rubber, Hemp, Soapstone and Empire Packing,**

Cordage,    Waste,    Wire Rope.
Oils,               Leads,
    Paints,           Glass.

Earl's Steam Pumps, Cameron's Steam Pumps, Pulsometer or Magee Pump.

Fairbank's Scales,    Hall's Safes,
Fire Brick,        Fire Clay,
Salamander Filling,    &c., &c., &c.

**Cleveland Dock,**        **Marquette, Mich.**

---

That fall, Terrence Moore purchased the machinery for a reduction mill from Fraser, Chalmers and Company of Chicago. It was late December before the mill equipment was finally delivered to Ontonagon aboard the steamer *Manistee*. There it was loaded aboard scows and towed to Iron River by the Ontonagon company's tug *Grace*, then hauled by sled up the snow-covered wagon road to the mill site. The site selected for the mill was on the bank of the Iron River on the Superior Silver Mining Company's property. A substantial wooden building was erected during the winter to house the machinery. The mill equipment was installed by the manufacturer's representative in May 1875, and a cargo of mercury, used in the processing, arrived aboard the first steamer into Ontonagon that spring. All the company needed now was an experienced hand to operate the mill. The Iron River Silver Reduction Works found their superintendent in the person of F.W. Crosby, who came to Iron River from the Judd & Crosby Reduction Works of Georgetown, Colorado, to take charge of the mill.

Finally, on July 2, 1875, the mill started running, processing ore that had been raised during shaft sinking at the Ontonagon mine. Immediately upon startup, some problems with the mill equipment became apparent. The mill had been built during the winter while the ground was frozen, and the foundation for the stamp mill was not solid enough to stand up to the constant pounding of the stamps. The mill run had to be stopped frequently to adjust the stamp mill, which tended to shift on its soft foundation, so the plant, which was designed to treat 10 tons per day, only averaged 1 ton per day over the three-week run.

The silver-ore processing plant was composed of a Blake jaw crusher, a Gates five-head stamp mill, two Varney amalgamating pans, a settling tank and a retort (distiller) for separating the silver from the mercury. Power for the mill was provided by a steam engine. Rock from the mine was broken down to walnut-sized pieces in the jaw crusher, then fed into the Cornish stamp mill. In the mill, the ore was pulverized by heavy iron stamp heads that were alternately raised by cams and allowed to drop onto the ore in the cast iron mortar. The ground ore was washed out of the mill through punched metal screens and into one of the amalgamating pans by a stream of water. The pan, which held more than half a ton of coarsely pulverized ore, consisted of a rotating iron plate that ground the ore against the heavy iron bottom of the pan.

After about five hours of grinding, the ore was reduced to a fine paste. Mercury was then poured into the pan, and the ore was mixed with the mercury for several more hours. This continuous agitation thoroughly mixed the metallic silver with the mercury to form an amalgam. Once the silver had been taken up into the amalgam, the pan was emptied into a settling tank. Any amalgam left was scraped out of the pan, leaving it ready for its next load of ore. In the settling tank the silver/mercury amalgam was allowed to settle to the bottom while the lighter waste rock was washed away. The silver-bearing amalgam was distilled in a retort, heating it hot enough to boil off the mercury while leaving the silver behind. The mercury was condensed and saved for reuse, while the silver was melted in a crucible and cast into ingots or bricks.

At the end of the first test run, 532 ounces of silver was recovered, a yield of $33.28 in silver per ton of ore. Superintendent Crosby estimated that another $300 worth of amalgam had been "absorbed" by the pans – the amalgam was filling every corner and seam in the machine – and that subsequent runs would show greater returns. The silver from the mill run was cast into a brick and taken first to Marquette and then to Chicago to display to the mine owners – and potential investors!

After reinforcing the stamp mill foundation, Crosby ran a test of ore from the Scranton mine. The ore had to be hauled nearly five miles to the mill, first by rough trails down to the mouth of the Iron River, then back up the wagon road to the mill. The 4½ tons processed yielded $203.75, an average of $45.28 per ton.

A third test run was made of 5 tons from the Superior silver mine in August. The 224 ounces of silver produced – a yield of $56 per ton – was cast into a 5¾-by-3-by-2½-inch brick, solid evidence of the value of the Iron River silver veins.

With successful mill tests of silver ore from three different mines, and a slight easing of the economic depression from the panic of 1873, the Iron River district was becoming a lively area. Captain Beaser hired surveyor C.H. Pratt of Ashland to lay out the town of Silver City on his land at the mouth of the Iron River. The broad streets and avenues were named for prominent figures from the Civil War, such as Farragut, Seward, Sheridan, McClellan and, of course, Lincoln. The busy main street along the river was called Beaser Street. Forty building lots had been spoken for by the time the plat was completed. The Ashland Lumber Company shipped in a large consignment of building supplies for the new town. Dr. Edmund Krelwitz of Humboldt, the president of the Porcupine Silver Mining Company, built a new building and opened a store. Mellen and Shelden of Houghton fitted out a former warehouse as a store and went into business supplying the mines and the new village. C.G. Collins of Marquette showed his confidence in the new silver district by closing his store in Marquette and moving his entire stock to Silver City. New roads were cut and graded to the mines, and a stagecoach line inaugurated daily service between Ontonagon and the boom town. In December 1875, a visitor described Silver City as being "in full bloom – six new buildings have been erected, and the town is alive with carpenters, store keepers, explorers, dogs and Indians, and it is expected soon to rival many of the larger mining cities on the lake."

Numerous new silver mining companies were started up in 1875: the Atlas, the Negaunee, the Wolverine, the Silver Mountain and the Mohawk silver mining companies of Negaunee; the Ishpeming Silver Mining Company of Ishpeming; the Eureka Silver Mining Company of downstate Greenville; the Luzerne Silver Mining Company (with investors from Menominee, Michigan, and Luzerne County, Pennsylvania); and the New York, Cariboo and Centennial silver mining companies of Marquette. Many of these companies had no more to work with than a parcel of land in the general vicinity of the Iron River silver finds and a large dose of optimism. Most of the new companies spent much of their meager capital paying mineral experts to prospect their lands, but the Porcupine Silver Mining Company made a unique offer. The company, formed by the St. Clair brothers and other iron mining men from Humboldt, offered 5,000 shares of the company's stock to anyone who discovered a silver vein on the company's land.

---

**A. B. MEEKER & CO.,**

**Pig Iron, Iron Ore,**

**STEEL AND IRON RAILS,**

93 Dearborn St., CHICAGO.

Sole Agents for

**Joliet Iron and Steel Co.,**

**National Iron Co.,**
De Pere, Wis.,

**Menominee Iron Co.,**
Menominee, Mich.,

**Green Bay Iron Co.,**
Green Bay, Wis.

Following the test runs of ore from the Ontonagon, Scranton and Superior silver mines, the mill sat idle for a month until a second test of Superior mine ore was made in September 1875. Captain John Spaulding was hoping to sell the E ½ of the SE ¼ of Section 13, just across the section line from the Superior, to a group of investors from Cleveland. To prove the value of the vein that ran across the two properties, Spaulding contracted with W.E. Westbrook to clean out the clay and water from the Superior adit and to mine 5 tons of ore from the full width of the vein. The investors supervised the mining to ensure that no selection of high-grade ore took place. The ore was crushed, and a 220-pound sample was randomly selected and sent to Cleveland for independent assay. The remainder of the ore was run through the mill under the scrutiny of the Cleveland investors. Samples were taken and assayed at every step of the process. The ore run through the mill produced $41.07 in silver per ton, but the assays of the various test samples were never made public.

The next mill run was on a small lot of ore from a new silver vein far south of the known Iron River veins. The Pittsburgh Silver Company had been incorporated in Marquette to work a vein of silver-bearing ore that had been discovered by Dr. A.S. Amerman of Champion and John G. Truax of Marquette in 1873. The find was on the W ½ of the SW ¼ of Section 34 T50N-R42W. Amerman was president of the new company, and Truax was secretary and treasurer. They were joined by mine and quarry owner W.L. Wetmore of Marquette, John R. Case of Champion and blast-furnace owner James Pierpoint of Pennsylvania. The company sank an adit 20 feet deep on the vein and mined about 5 tons of ore. It then cut a rough horse trail some 10 miles to the mill and packed out 920 pounds of ore for a test run. Despite assays showing the ore to be worth up to $119 per ton, this bulk sample of what was

---

CHAS. E. WRIGHT,

## IRON EXPERT,

Will attend to the Examination of

### Ores, Mineral Lands,

AND

### FURNACES.

Reference Furnished on Application.

———

Office and Laboratory in Maj. Brooks' rooms, Adams Block, third floor, Front St., Marquette, Mich.

undoubtedly selected ore yielded only $27.38 per ton in the mill and produced a mere 10-ounce ingot. No more ore was mined from the Pittsburgh prospect, but the company was back in the news a few months later. When it sent the ingot to the mint to be refined, $6 worth of gold was recovered along with the silver. This was the only gold reported from the Iron River Silver Range despite numerous tests and assays, raising the suspicion that the results were fraudulent. Perhaps the vein rock had been salted, or maybe somebody had thrown a few gold coins into the process.

   Another company organized in Marquette in 1875 was the Cleveland Silver Mining Company, which bought 80 acres from C.G. Collins just south of the Ontonagon and Lake Superior Silver Mining Company's mine. The new company's explorer, W.E. Westbrook, traced the silver vein from where it made a turn on the Ontonagon and Lake Superior property and followed it onto the Cleveland company's parcel. The new company had solid backing, its stockholders being some of Marquette's founding fathers and iron ore capitalists. Judge J.W. Edwards of Marquette was president, and mining hardware merchant Colonel James Pickands was secretary. Other investors included Marquette pioneer settler and alderman Amos R. Harlow, railroad agent A.A. Ripka, iron mining agent Jay C. Morse and steel maker A.B. Meeker of Chicago. The company sent Burr Hursley to the Iron River district to begin development of its silver vein. A test shaft on the silver-bearing formation on the company's land showed the vein to be only 8 to 10 inches wide, but very richly charged with silver.

   Charles E. Wright, the assayer and analytical chemist from Marquette, visited the new Cleveland Silver Mining Company's shaft on the way to his new position with the Wisconsin Geological Survey. He estimated the vein rock to be worth $200 per ton in native silver. He also clarified a misconception about the silver-bearing formation. Mining men had been referring to it as a fissure vein, since that was the familiar formation in which precious metals were commonly found. Professor Wright found the Iron River silver "vein" to actually be a layer of partially metamorphosed sedimentary rock. He theorized

Stamp mill, amalgamating pans and settling pans as used at the Iron River Reduction Works.

that this stratum could be found in the same stratigraphic position between the slate and the sandstone through much of the county.

The Cleveland Silver Mining Company announced its plan to finance the development of their prospect by levying an assessment of 10 cents per share every month for five months. With the first assessment paid in, it immediately set a force of 20 men to work under Burr Hursley's supervision. Within a few weeks, they had erected three log buildings and had started a new production shaft. With permanent accommodations for the miners in place, the Cleveland company was prepared to continue development work through the winter. Mining Captain Thomas Mellen took over management of the shaft sinking and underground work, while Hursley went to work for the Ontonagon and Superior companies, developing the surface facilities at their mines.

At the Ontonagon and Superior mines there was renewed activity. Management of the two companies had been turned over to Samuel P. Ely, the longtime manager of several successful iron mines in Marquette County. Ely brought his personal wealth as well as his experience to the job, paying off the companies' outstanding debts. This influx of capital allowed the Ontonagon company to buy out the Scranton Silver Mining Company and gain full ownership of the reduction works. Ely also persuaded Captain Gilbert D. Johnson to take on the position of superintendent at the mines. Johnson was one of the pioneer iron miners at Ishpeming, had been the first president of that newly incorporated village in 1869 and had served as the superintendent of the Lake Superior iron mine for 20 years. Mill Superintendent Crosby had resigned in September, so the company hired mining engineer A. Heberlein to take over his position. Heberlein inspected the mill and found a broken stamp head as well as a leaking amalgamation pan. The company also ordered another five-head stamp mill along with a second pair of amalgamation pans from Fraser and Chalmers of Chicago to increase the mill's capacity to 20 tons per day.

The underground workings at the Ontonagon and Superior mines were pumped out and mining of the silver ore resumed. The mill was put to work on ore from the two mines and was kept running nearly full time. Due to the broken stamp head and the soft stamp mill foundation, the mill still only averaged 4 tons per day. The Superior was 51 feet deep, and a whim (a horse-powered hoisting windlass) had been erected to hoist ore up the inclined shaft. The Ontonagon's inclined shaft was down 100 feet, and drifts had been extended 60 feet on each side of the shaft. A total of 63 men were employed at the two mines: seven men in the mill, 24 miners, 17 laborers, six carpenters, two masons and three teamsters, as well as the superintendent, captain, doctor and clerk.

At the Cleveland Silver Mining Company's property, 30 men were at work in December. A superintendent, 16 miners, four teamsters, six laborers, two carpenters and a blacksmith made up the workforce. The No. 1 shaft had been sunk at an angle and struck the silver vein in mid-December 1875. The silver bearing lode was found to be 16 inches wide and rich in silver. While most of the silver in the Iron River range had been found in the form of small flakes, the Cleveland mine produced specimens carrying thin sheets of silver. As the miners followed the vein deeper over the next few months, the silver ore grew richer, showing thicker sheets and occasional nuggets of silver. One section of the vein yielded a three-quarter pound nugget, while other pieces from one-half to 1½ ounces were found. The ore was valuable enough that visitors were only allowed into the workings in the company of Captain Mellen.

At the Collins mine, superintendent P.C. Beanston was working a crew consisted of a foreman, six miners, six laborers, two carpenters and one blacksmith. The location had five sturdy buildings – an office, bunk house, cook shack, shop and barn. Mining Captain Dan Powell was in charge of underground work. The No. 1 shaft was down 55 feet by January 1876, and a second shaft was started 375 feet away. The company planned to open up stoping ground by drifting between the two shafts and hoisting ore through both of them. Shaft sinking alone was producing a ton of ore a day from the 13-inch vein, along with 4 tons of waste rock. The ore was stockpiled in anticipation of processing it in the Ontonagon company's reduction works.

The machinery to expand the mill arrived on the silver range in December 1875. The manufacturer sent its expert machinist, Mr. Myrick, to install the new machinery and to overhaul the original equipment. The foundation for the first stamp mill was rebuilt, and the broken stamp head and leaking amalgamation pan were repaired. After a three-week shutdown, the mill was restarted, once again running on ore from the Ontonagon mine. Even with the new and refurbished equipment, the recovery of silver from the ore was disappointing, however. Experts estimated that anywhere from one-quarter to one-half of the precious metal in the ore was lost in the tailings, along with a significant portion of the mercury. It seems that the fine, often microscopic particles of silver were not recovered efficiently by the stamping and pan amalgamation process.

The poor recovery of silver in the mill, along with a

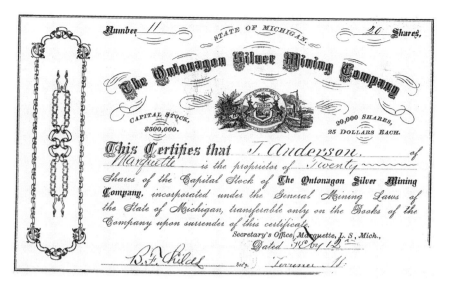

streak of low-grade ore in the mine forced the Ontonagon Silver Mining Company to close down its operations for the winter in mid-January 1876. There simply wasn't enough silver coming out of the mine and plant to cover expenses or to pay back Mr. Ely for his investment in the mine. The stockholders were faced with the choice of issuing more stock and trying to sell it to raise capital, taking out a mortgage on the property or selling the mine outright. The appointed date for a special meeting of the stockholders to consider these options came and went, but no word of their decision was published, and the Ontonagon Silver Mining Company faded from the public's attention.

    Meanwhile, the Cleveland Silver Mining Company continued the development work in its mine. Captain Mellen started a second shaft about 400 feet away, sinking it vertically to strike the vein at depth. Shaft sinking continued through the winter of 1875-76, finally reaching the vein in April at a depth of 136 feet. The silver-bearing quartzite and slate totaled 40 inches at the bottom of the shaft, and drifts were started along the vein to develop the mine for production.

    Despite the Ontonagon company's troubles, the Scranton Silver Mining Company still planned to reopen its prospect in the spring and took out a lease on the mill, planning to mine and process its own ore. The other companies were not so confident that the Ontonagon company's reduction plant offered an economic method of recovering silver from the Iron River veins. In February, the Ontonagon and Lake Superior Silver Mining Company sent a 1-ton sample from the Collins mine to the Chicago Reduction Works. The ore yielded $40 per ton. Financially strapped and discouraged by the mill test, the company closed down its operations in the early summer of 1876, leaving a stockpile of silver ore alongside the abandoned shafts.

The Cleveland Silver Mining Company also opted to try alternatives to the stamping and amalgamation process. Rather than running its ore through the Ontonagon company's stamp mill and pans, the Cleveland Silver Mining Company decided to ship 10 tons to the Silver Islet mine on Lake Superior's north shore to be concentrated in Captain William Bell Frue's invention, the Frue vanner, a vibrating wash system for separating out the ores. While the Silver Islet plant's superintendent was said to be most satisfied with the test, he did admit that the process was not ideal for the Iron River silver ore. The silver concentrate was shipped to the Wyandotte Works to be refined and cast into bullion, but the final results of the test were never made public.

After a visit to the Cleveland mine in the summer of 1876, company president J.W. Edwards and Captain Mellen were returning to Marquette. It would be a day or two before the next scheduled passenger steamer arrived, so they booked passage on a coastal freighter that was at the dock in Ontonagon. The *St. Clair* was an old wooden steam barge that was bound from Duluth to Marquette by way of Bayfield and Ashland, Wisconsin, then on to Ontonagon and Houghton, carrying a cargo of flour, grain and livestock. She also had spartan accommodations for passengers. The boat left Ontonagon just before midnight on July 8, 1876, and was two hours into her voyage and about seven miles offshore when fire was discovered near the boiler room. The captain ordered the blazing vessel turned toward shore and attempted to launch the large yawl boat, but the crew were driven back by the flames, which quickly destroyed the yawl. A small metal lifeboat on the forward deck was launched and quickly filled with a dozen passengers. One last passenger trying to jump into the boat capsized it, throwing all of them into the chilly water.

Once they had righted it, some were able to get back into the swamped lifeboat, only to have it capsize again and again. One by one those who had escaped drowning outright began to die of exposure. Of the 16 passengers and 15 crew, only five ultimately survived, paddling the lifeboat to shore with planks and seat boards. Among the dead were Captain Mellen and Judge Edwards. Mellen's body was recovered the next day, but Edwards was never found.

The Cleveland Silver Mining Company was the only mining company still operating on the Iron River range in the summer of 1876, and in September it became the last to close down. The loss of the company's president and mining captain probably had some bearing on the closure, but what really killed the Cleveland and the rest of the silver mining companies was a combination of economy and geology.

The Panic of 1873 had thrown the United States into the worst depression the country had ever experienced, one which lasted nearly 10 years. There simply wasn't enough capital available in those tough economic times to bankroll the companies through the long, costly evolution from prospect to paying mine. Underground workings had to be developed to the point where sufficient quantities of ore could be mined efficiently; and processing plants of a large enough size to process the ore economically needed to be built. None of this could be done without large expenditures before any returns could be expected from the mines.

The mining companies' other nemesis was the character of the silver ore in the Iron River district. Despite a few rare specimens of silver in sheets and nuggets, the silver in the ore was predominantly found in tiny flakes. The Cornish stamp mills didn't grind the ore

Captain Thomas Hooper

fine enough to liberate all of the microscopic silver from the rock, and tiny particles of silver were often carried away with the tailings. In addition, the settling pans used to separate the mercury/silver amalgam from the poor rock often discharged amalgam along with the tailings if not operated with great care and skill.

Despite the failure of all of the silver mining companies and the difficulties encountered in trying to extract the silver, further attempts were made to mine the Iron River district in the 1880s. The Milwaukee and Lake Superior Silver Mining Company was incorporated in Wisconsin in the fall of 1882, but was unable to raise enough capital to start operating until two years later. Organizers of the company included H.N. Smith and C.W. Brown of Milwaukee, and William Butler and Charles M. Howell of Ontonagon.

Austin Corser
AL WEATHERBURN

The new company, backed by investors from Chicago, bought the Ontonagon and Superior silver mines. It assayed ore from the shaft and adit, finding it as rich as previous reports had stated. The company also sampled the tailings and found that up to $263 in silver per ton had been lost in this waste from the mill.

Realizing that the old reduction mill was not an effective method of recovering the silver, the new company shipped a test lot of ore to a smelter in Chicago and also sent 24 barrels to an experimental smelter run by Daniel Merritt and Professor D.F. Henry at the Iron Bay Foundry in Marquette. Apparently neither test was satisfactory, since the Milwaukee and Lake Superior Silver Mining Company abandoned the mines for another two years.

In 1887, the Milwaukee company leased the Ontonagon mine to another group of Chicago investors. The new operators sent ore to Chicago for assay and to a custom mill in the east for a mill test. Once again the assays were promising, so they overhauled the Iron River reduction plant and put it back into operation, only to shut it down a few months later, ostensibly for the winter, but, in fact, for good. There were further attempts to reopen the Iron River silver mines in 1888 with a new group of Chicago investors, but no more silver bullion was ever produced.

The problem of recovering the fine particles of native metal wasn't unique to the silver ores; the same difficulties plagued the many copper mines opened in the Iron River district. Like the silver, the copper in this area was found as fine particles in sedimentary rock.

Of the dozen or more copper mines opened in the Porcupine Mountain/Iron River area in the 1800s, none were able to produce copper at a profit until Captain Thomas Hooper took over the Nonesuch copper mine. After the silver mines closed, Hooper found himself without a job, so he leased the Nonesuch, which was about four miles southwest of the silver mines. The mine had been in operation since 1867 and had never turned a profit. By carefully selecting only the richest ore in the mine before it was raised and sent to the mill, Hooper was able to produce a few tons of copper at a profit for three years before selling the mine.

It wasn't until World War I and its attendant inflated prices that another copper mine in the district was able to make a profit – this time the White Pine Mine three miles south of the silver mines. The mill there only recovered about two-thirds of the copper in the rock, leaving one-third of the metal in the tailings – no better than what the silver mill had been able to achieve 40 years earlier.

The White Pine mine produced not only copper, but also silver – about 82,000 ounces of it in 1916. The end of the war spelled the end of the White Pine mine until the mid-1950s. In the meantime, methods of froth flotation of copper ore had been developed that enabled modern miners to recover a much greater proportion of the metal. The mine was reopened, along with a new mill with flotation tanks to recover nearly all of the desired minerals in the ore – chalcocite, native copper, and silver. The White Pine mine continued to produce silver as a byproduct of the copper, approximately 1 ounce of silver for every 100 pounds of copper, until it closed in 1995.

# Chapter 4

 ## The Ropes Gold Mine

Traces of gold had been found in Michigan since the middle of the 19th century. The Houghton brothers' finds in the 1840s, the gold found on the silver lead range in the 1860s, Waterman Palmer's discovery in 1865 – all of these showed that the precious metal existed in the state, but could it be mined profitably?

It was Julius Ropes who finally persuaded the world that Michigan had gold in economic quantities. Born in 1835 in Newbury, Vermont, and educated at the St. Johnsbury Academy, Ropes came to Marquette County in 1855. Ropes worked in a store in Harvey, which was then known as Chocolay, delivering mail, hauling it by dogsled in the winter. He later joined Marquette's first mayor, Henry H. Stafford, in running a drugstore in Marquette. In 1867, Ropes moved to Ishpeming and opened a pharmacy for Stafford.

Their store was the second business in the young town, the first being a saloon (although some histories suggest it was actually a brothel).

Ropes soon bought out Stafford's interest in the store, going into business on his own as J. Ropes & Company. He married Silas Smith's niece, Eunice Louisa Rouse, in 1868, was a member of the first school board, and was appointed postmaster. Ropes recognized the potential for precious metal mining in Marquette County and assisted the mineral prospectors by performing fire assays of gold and silver ores, starting during the silver lead boom of 1863-64. In 1878, he sold his pharmacy business to F.P. Tillson, but continued his assay work from an office on

Main Street, above his brother-in-law's shoe store.

In the late 1870s, woodcutters chopping wood for the charcoal kilns of the Deer Lake Furnace brought Ropes specimens of what they thought was petrified wood. Ropes identified the mineral as a type of asbestos. He explored the area where the "petrified wood" had been discovered and found a range of hills with outcroppings of a green serpentine marble, which he recognized as having commercial value. Ropes bought the mineral rights to the S ½ of the NW ¼ of Section 29, T48N-R27W in 1879. That year, Ropes, along with Dr. William T. Carpenter, Solomon S. Curry, and George P. Cummings, a fellow Vermonter who was married to Ropes' sister, Hannah, founded the Huronian Marble Company. Aside from cutting and polishing a number of specimens, Ropes did little with the property, but continued to explore, realizing that the geology of the serpentine range was favorable for precious metals. He soon found a small vein of quartz that looked promising for silver. Upon testing the ore in his laboratory, he found that it contained both gold and silver.

On May 17, 1881, Ropes stuck his prospecting pick into a mossy rock outcrop and revealed a vein of quartz containing $21 per ton in gold, enough for commercial exploitation.

Ropes sank an 11-foot shaft on this vein and found that the lode widened from a few inches at surface to 2 feet at the bottom. Samples of gold-bearing quartz from the shaft were found to be worth up to $367 per ton.

Julius Ropes
SUPERIOR VIEW

In August, the Ropes Gold and Silver Company was formed, with Ropes, Cummings, Curry and Carpenter as partners. Capital stock was $1 million, consisting of 40,000 shares with a par value of $25 each. The company bought the eastern half of the Huronian company's land for $5,000. During the winter of 1881-82, the company sank a second shaft known as the "B" shaft about a quarter-mile west of the "A" shaft at the discovery point. Assays of specimens from this shaft ranged from $18 to $59 per ton in gold and silver. A series of trenches dug across the strike of the vein traced it several hundred feet farther west onto what was still the Huronian company's land.

At the March 12, 1883, stockholders' meeting, it was decided to purchase the rest of the Huronian Marble Company's land for $20,000 and to sink an 8-by-10-foot main shaft at the vein's richest spot. This shaft, named the Curry shaft after mine superintendent Solomon S. Curry, eventually became the only production shaft and ultimately reached a depth of 813 feet. By late summer, the new shaft had reached a depth of 35 feet. It was equipped with an open "gallows" style headframe and a horse whim for hoisting. A steam-powered pump had been installed to keep the workings dry, and a barbed-wire fence had been added to keep the curious at a distance. A large dog named Bouncer assisted in the guard duties.

In August 1883, a small Fraser and Chalmers stamp mill was installed, and the first batch of ore was processed. After a run of 39 days, the company recovered a total of 40 ounces of gold, worth $830, and 158 ounces of silver worth $173, from 100 tons of rock. Local newspapers heralded the production as "A Golden Bonanza!" and the first "bullion train" (actually a horse-drawn wagon) in Michigan carried the gold from the mine to Ishpeming for public display. The Ropes mill made two more runs totaling 172 tons, yielding $820

Solomon S. Curry

ALL IMAGES AUTHOR'S COLLECTION UNLESS INDICATED

Curry shaft house, Ely shaft and 1887 mill building
MARQUETTE COUNTY HISTORICAL SOCIETY

in gold and $176 in silver, before shutting down for the winter. The bullion was shipped to the Philadelphia Mint for refining. When the first coins minted from Michigan gold were received in December, they were eagerly sought as collector's items.

Underground development work and mining resumed in April 1884. A steam-powered hoisting plant formerly used at the Dexter iron mine, good for a depth of 300 feet, was installed in a new building at the shaft. The mill reopened in June, financed by the sale of 24,000 shares of stock. Over the summer the small mill stamped some 163 tons of ore, yielding $2,613 in gold and silver. In the fall of 1884, the company decided to enlarge the mill from five to 25 stamps, and the expanded mill, housed in a new 56-by-76-foot, three-story building, went into production in November of the same year. The new mill building was connected to the shaft house by a trestle. Tram cars were loaded at the shaft house and pushed across the trestle to the top floor of the mill, where the ore was dumped onto a coarse screen called a grizzly. The grizzly separated the large chunks of ore, which were crushed in a Blake jaw crusher, from the fine ore that bypassed the crusher and was shoveled into the Fraser and Chalmers Cornish stamp mills.

The Cornish stamp mill, named for Cornwall, England, where it was developed, consisted of five cast-iron stamp heads or pestles, which were alternately raised and dropped onto an iron die in a common mortar.

Frue vanners in the 1884 mill
MARQUETTE COUNTY HISTORICAL SOCIETY

Five such mills, or stamp batteries, were installed at the Ropes. The cylindrical stamp heads, each weighing 750 pounds, were raised eight inches by a camshaft turned by steam power, then allowed to drop 66 times per minute, crushing the coarse ore in the mortar, which was kept nearly full of mercury. In addition to crushing the ore and freeing the gold from the rock, the stamps also brought the particles of gold into close contact with the mercury, forming an amalgam – a soft alloy of gold, silver and mercury. Punched metal screens across the outlet of the mortar kept the ore and amalgam in the mill until it was finely ground. A continuous flow of water washed the fine ore and amalgam out through the screens and onto the amalgam-collecting plates. These inclined 48-by-120-inch sheets of silver-plated copper would catch the metallic amalgam, while the waste rock was washed through a trough to the vanner room, 10 feet below and to the north of the stamp room.

After a run of two weeks to a month, depending on the richness of the ore, the flow of ore to the stamp battery would be halted and the mill stopped for cleanup. All of the gold mercury amalgam would be removed from the mortar and plates and retorted (distilled), boiling off the mercury and condensing it for reuse, leaving the gold behind. The mortar was then refilled with mercury and another run was begun.

A great deal of the gold and silver in the Ropes ore could not be recovered by mercury amalgamation, because it was locked up in sulfide minerals, including pyrite, chalcopyrite, galena and tetrahedrite. These values were recovered from the stamp mill tailings by gravity separation in Frue vanners. The Frue vanner, developed by William Bell Frue, the superintendent of the famous Silver Islet mine off the Sibley Peninsula in Ontario, was an endless rubber belt which traveled up a slight incline at six feet per minute while vibrating from side to side at 200 strokes per minute. The stamp mill tailings were fed in a slurry onto the belt near the top of its slope. The heavy mineral-bearing pyrite settled on the belt and was carried over the upper end, while the lighter rock was

1887 mill building
MARQUETTE COUNTY
HISTORICAL SOCIETY

kept in suspension by the vibration and washed off the lower end. This waste rock was dumped into the valley to the north of the mill, while the mineral-rich concentrate was sent to the Aurora Smelting and Refining Company in Aurora, Illinois, for refining.

Inside its wooden building, the mill was able to run even in subzero winter temperatures. By the end of 1884, the shaft was 85 feet deep and producing on two levels. Other improvements were made at the site that year, which included adding a boarding house for the miners, carpenter and blacksmith shops, an assay office, a warehouse and a barn. The stream in the valley just to the north was dammed with an earthen dam reinforced with tailings to increase the water supply to the new mill. Mining and milling continued on two shifts until October 1885, when the night shift was discontinued in an effort to conserve the scarce working capital. By that time, the shaft had reached 250 feet with four levels. The levels were usually spaced about 50 feet apart vertically, but sometimes varied depending on mining conditions.

Superintendent George Weatherston
KAY WALLACE PORTER

1886 saw the introduction of power drills at the mine, financed by a 10-cent-per-share assessment. Powered by air from a steam driven compressor, the four Rand "Little Giant" drills allowed the force of miners to be cut from 30 to 12 and permitted drifting (driving the horizontal underground passages) to be carried out at a rate of 3½ feet per day. During July, a prolonged drought dried up the mill's water supply, so another dam was built at the junction of three small streams about a mile to the west and a pipeline was run to the mill.

A total of nearly 50 men were now working in the mine and mill, enjoying the new steam-heated dry house, complete with hot and cold running water. The mine was opened down to the sixth level, with an estimated 38,000 tons of ore available for stoping. As the shaft passed the 300-foot mark, the hoisting plant became inadequate, and a new 6-foot hoist drum capable of hoisting 1,200 pounds of ore was purchased from Marquette's Iron Bay Manufacturing Company.

Although the mine was 349 feet deep and producing from five of its six levels, by 1887 the lack of capital had taken its toll. Worn-out equipment was not being maintained and the mill building itself was falling into disrepair. When Frank Cummings, an experienced gold mining man from Colorado, took over from George Weatherston as superintendent in May, he immediately started a program of repair and renovation. He began by replacing a cracked stamp-mill mortar, which was leaking precious gold mercury amalgam onto the floor. When the floor in the mill was replaced, the workers were actually able to scrape amalgam from the underlying timbers. Cummings was responsible for other improvements at the mine, including the installation of riffles in the tailrace to save gold from the tailings. He rerouted the oily steam engine exhaust, which had been interfering with the amalgamation process, away from the mill's water supply. He also introduced the six-day work week – a reduction from the usual seven days a week – feeling that "one day's rest each week harms no one." After three months, Cummings resigned, apparently in frustration over the lack of money needed to put the operation in proper order.

Stamp mill and amalgamating plates in 1887 mill

Further improvements were begun in 1887. It was decided to build a new mill building with room for 40 stamps, but initially equipped with 20. A 6-foot-high wooden dam was built on the Carp River about three-quarters of a mile from the mine, with a 20-inch Victor turbine driving a 5½-inch duplex Knowles pump to ensure a constant water supply to the mill. A new shaft, named the Ely for company secretary Clarence R. Ely, was started 350 feet east of the Curry shaft and sunk 75 feet during the year. To finance these improvements, another assessment of 50 cents per share was called in August.

The new mill began operation on May 17, 1888, adding 20 stamps to the existing 25, all running off a new 235-horsepower Corliss steam engine. With the new equipment nearly doubling the mill's capacity, the small, crooked shaft became a bottleneck, prompting the company to square up the shaft and to install a skip road (wooden rails on which the skip or ore car would ride). Down to the eighth level, the shaft had followed the 80-degree southerly dip of the vein. It was decided that in order to increase hoisting efficiency, the shaft would be sunk vertically below that level, even though this would carry the shaft away from the gold-rich quartz vein.

Installing the skip road to the fifth level necessitated shutting down the mine and mill for more than a month in October and November, during which time a new shaft house was built and a new 30-ton-per-hour Gates gyratory crusher was installed, replacing the outmoded Blake jaw crusher. The old crusher had been located inside the old mill building, but the dust it produced caused excessive wear on the mill machinery. The new crusher, along with its own 50-horsepower engine, was installed in the shaft house.

A possible solution to the mining company's continuing capital shortage was proposed in 1888 by General Russell A. Alger, who led a group of Detroit investors in an offer to buy a majority interest in the Ropes Gold and Silver Company. The stockholders rejected the offer, feeling that the $3 per share offered was too low, and another 50-cent-per-share assessment was called in September.

By the end of the year, the Curry shaft was 441 feet deep with nine levels. A total of 62 men found employment at the mine in 1888: 40 miners and other underground workers, 15 men in the mill and seven surface employees.

**A Ropes crew**
KAY WALLACE PORTER

With reserves of high-grade ore still in sight, the company decided to complete its expansion in 1889 by installing another 20 heads of stamps, for a total of 65 heads in the two mill buildings, and four more Frue vanners. The expanded mill went into operation on March 9. A larger air compressor capable of powering up to 12 drills was also added. The shaft was extended to the 11th level at a depth of 551 feet, and a Lane band friction hoist was installed, powered by the old stamp mill engine. The enlarged, improved mill, along with other cost-cutting measures, trimmed mining and milling overhead to $1.75 per ton. The gross production of $99,715.45 from 31,365 tons, the highest in the mine's history, left the company more than $6,000 in the black at the end of 1889.

During 1890, the Ropes company contracted with M.E. Harrington and Son of Ishpeming to search for additional quartz lenses with diamond drill. This was the first underground diamond-drill exploration on the Ishpeming Gold Range. The exploration program paid

**Ropes mills**
MARQUETTE COUNTY
HISTORICAL SOCIETY

off almost immediately, with a vein of low-grade ore being found north of the fifth level. This vein was mined by a crosscut from the Ely shaft. Another vein, assaying $60 per ton, was found 70 feet north of the 11th level later that summer. A drift was run toward this vein from the 10th level, but when it reached the vein, the quartz was only an inch wide. Fortunately, the lens widened out to 5 to 6 feet as it was traced to the 11th level.

The mine was still in good shape at the end of 1890, the shaft being 639 feet deep with the skip road extending to the lowest level, the 12th. The previous year's profits had been spent, however, and another 25-cent-per-share assessment was called in February 1891.

The mine operated quietly the next year, sinking the shaft to the 13th level. The old 25-stamp mill was shut down, leaving only the 40 heads running in the new mill building. The air compressor was enlarged to a duplex (two cylinder), doubling the number of power drills which could be used in the mine. Pay rock was mined from the fifth through the 13th levels.

Rich rock from an ore lens on the eighth and ninth levels, 200 feet from the shaft, allowed the mine to make a profit during the first two months of 1892, but by July a 25-cent-per-share assessment was called to ensure a supply of reserve capital to keep the mine open. A rich streak assaying up to $213 per ton was found the next month, but any profit from this was offset by the expense of drifting through worthless or low-grade rock to get to the pay rock.

By December, when the night shift was cut out to reduce costs, rumors were being heard that the mine would close, but the reduced hours and the cessation of all development work, mining only the rock already exposed, allowed the mine's production to exceed expenses during the last months of the year. At the end of 1892, the shaft had reached 813 feet, 25 feet below the 15th level, never to go deeper. Sinking vertically had carried the shaft away from the vein, necessitating costly drifting through poor rock. The financial crisis that became known as the Panic of 1893 curtailed investors' ability to put needed capital into their businesses, so at the March 1893 annual meeting, the superintendent recommended that further sinking be done in the rich ore body itself. Thus, all rock removed would be pay rock, theoretically cutting expenses.

The 15th level was extended more than 300 feet to the east into a large body of pay rock, where an incline

**Ropes Gold Mine**
LAKE SUPERIOR MINING INSTITUTE

The vanner room in the 1887 mill
MARQUETTE COUNTY HISTORICAL SOCIETY

was sunk 100 feet into the ore body. The quartz vein here was 12 feet thick and less mixed with slate than almost anywhere else in the mine. It carried an average of $6 in gold but had specimens assaying up to $150. There was plenty of rock to supply the mill, but the added expense of hauling the ore up the incline and re-handling it at the 15th level ate up any profit, since extra workers and machinery were needed. The mine eventually reached its final depth of 915 feet below surface in an underhand stope at the bottom of the incline. Although the company explored on the surface for more veins, trenching across the strike of the vein west of the shafts and sinking a small exploratory shaft near the schoolhouse, most of the production over the next four years came from the bottom levels and from lenses of quartz previously exposed in the upper levels of the mine.

By 1897, the company's lack of working capital had finally caught up with it. A number of paydays had been missed over the last eight years, until the back pay due the miners totaled about $3,000. On July 10, 1897, the 40 miners stopped work and filed suit to get their pay. On July 27, the suit was decided in the miners' favor. Since the company did not have the cash to pay the judgment, the company's assets, including mine property, were turned over to a receiver, spelling the end of the Ropes Gold and Silver Company.

During its 14 years of operation, the mine had produced a total of $645,792 in gold and silver, with the best year's production being $99,715 in 1889. Although the company had operated in the black for several years, what little profit was made was put back into the mine. No dividends were ever paid to the stockholders; in fact, a total of $2.34 per share in assessments had been levied on the stock.

Some of the company's financial woes were caused by the inefficiency of the mill and poor recovery of the gold. The Ropes mill processed the ore by stamping it in Cornish stamp mills. The stamp mills would work very well on hard, brittle rock such as quartz, but the Ropes ore contained a great deal of soft talcose slate, which tended to stick in the mortar, cushioning the impact of the stamp head.

The company continually sought to improve the recovery of gold in the stamp mills. Early on, the original 60-mesh punched metal screens which controlled the outlet from the mortars were replaced with 40-mesh wire screens, improving the flow of ground ore from the mill. Experiments were made with the frequency at which the stamp heads fell, starting with the original 66 blows per minute and finally settling on 88. A consistent flow of crushed ore to the mill was the intent behind the installation of Tullock automatic feeders in the 25-stamp mill in 1886. These shaking-table feeders tended to clog, however, and were supplanted by Challenge disc feeders in the 40-stamp mill in 1888. Another improvement incorporated in the new mill was the use of 850-pound stamp heads in place of the 750-pound stamps in the original mill.

The company even considered replacing the stamp mills entirely, experimenting with the Wiswell Electric Ore Pulverizer, the Huntington centrifugal mill and the Crawford mill. The Wiswell Pulverizer was tried in 1885. This machine consisted of an 8½-foot circular crushing box with a V-shaped groove. Four heavy iron wheels,

Ropes mine and cyanide plant 1900
MICHIGAN TECH ARCHIVES

rotated by a central shaft, rolled in this groove. Ore and mercury were fed into the groove, where the rotating wheels crushed the ore. A large battery supplied electric current, which was passed through the pulp, supposedly assisting in amalgamation. At first the Wiswell machine seemed to be a success, recovering much more gold than the stamp mills and prompting the company to order a second pulverizer. When the bottom of the crushing box wore out, however, it began to seem like less of a bargain. The second Wiswell Pulverizer, with several improvements, was installed under the supervision of its inventor, but again wore out, costing $300 for repairs in a 46-day run. Both machines were abandoned, and the company continued to use the old, but reliable, Cornish stamps.

In 1890, the Ropes company ordered a 5-foot diameter Huntington centrifugal mill on approval. Rated at 15 tons per day, it was guaranteed to do the work of 10 stamps. Although this type of mill had been used with some success at the Michigan Gold Mine, it apparently didn't fulfill its promise at the Ropes, since it was never mentioned after being installed in July.

The major stockholders of the Ropes Gold and Silver Company were also involved in the Fire Centre Mining Company, which had several prospects north of the Dead River. When the Fire Centre prospects closed down in 1893, that company's Crawford centrifugal ball mill was brought to the Ropes mine and given a trial run. Like the rock of the Dead River range, the hard quartz from the Ropes quickly wore out the machine.

Much of the gold in the Ropes mine is found in association with pyrite, which inhibits the amalgamation of gold with mercury. The mill saved some of this ore by gravity concentration in Frue vanners, but shipping the gold-rich pyrite concentrate to Illinois for smelting cost $25 per ton, prompting the company to seek methods of treating it locally. Installation of a chlorination plant, in which the gold was released by roasting the ore and treating it with chlorine gas, was proposed in 1888, but never given a try. A smelting furnace was suggested several times and was tried on a small scale at the mine in 1890, using lead ores coming from the mine and from other local prospects. Apparently it was not economically feasible to operate a smelter locally, as the concentrate continued to be shipped "below" for treatment.

As was the case with most mines of that day, a settlement had sprung up at the Ropes Gold Mine location. At first the miners were housed in a log boarding house, but starting in 1883, the company built a double row of houses on the north side of the valley for the miners and their families. The residents were allowed free use of company land for pasturage, kept their own cows and planted gardens. In 1892, a company official proudly announced that the residents of the location had produced more than 1,000 bushels of potatoes. A two-story frame house for the superintendent was built on the hill overlooking the mine on the south.

A schoolhouse was built in 1886, and a schoolmaster and his wife, Mr. and Mrs. Irwin, joined the small community. Irwin was later replaced by Frank B. Wentworth. As part of a school district that included schools at the St. Lawrence iron mine and the Rock

Schoolhouse and miners' cabins at Ropes
MARQUETTE COUNTY HISTORICAL SOCIETY

Kilns, the Ropes Gold Mine Public School served students from kindergarten through high school. On Sundays, church services were held in the schoolhouse.

The mine location was linked to town by a road along the Carp River to the Deer Lake Furnace. At first, the road was little more than a rough trail cut through the woods, passable for heavy loads only when frozen solid in the winter, but improvements were soon made. The road was corduroyed with cedar logs and filled with waste rock from the mine and slag from the Deer Lake Furnace, and eventually became a county highway. Stage service to the mine was inaugurated in the winter of 1884. John Burke, who ran the City Livery and Boarding Stables in Ishpeming, started the service with a horse-drawn sleigh, making two trips a day. A round trip took an hour and cost 75 cents.

For the first four years of the mine's operation, no power tools were used, all drilling being done by hand. Miners would work in three-man teams, "double jacking," one man holding and turning the drill while the other two alternated hitting it with sledge hammers. Once a series of holes had been drilled, sticks of "giant powder" (as the new explosive we now call dynamite was known), were tamped into place and touched off. The broken rock was loaded into tram cars by pick and shovel and pushed by hand to the shaft where it was dumped into the skip. The shaft had been sunk 300 feet before power drills and the air compressor to run them were installed in 1886. Although this reduced the amount of hand labor, it also cost jobs, the force of miners being reduced from 30 to 12. Manual labor continued to be the only method of loading and tramming the ore, however.

The Ropes was considered to be a safe mine, and compiled an enviable safety record for its era. During the 14 years of its operation, only seven fatalities were recorded. A tragic record by today's standards, this was impressive when compared to the iron ore mines of that day, where several fatalities per year were not unusual. The gold mine was in hard rock and needed no timbers to support the drifts and crosscuts, while the iron ore mines of that period were mining soft, earthy hematite ores, which were prone to caving.

Accidents did occur at the Ropes, however, resulting from both falling rock and primitive working conditions. The first fatality happened on September 8, 1886, when miner John H. Martin was working in the 4th level,

Ropes Gold Mine
Public School
KAY WALLACE PORTER

Drilling blast holes on the 800-foot level of the Ropes Gold Mine
DAN FOUNTAIN

west of the shaft. The 44-year-old Cornwall native was shoveling in a low stope when a piece of rock weighing 2 tons fell, killing him instantly. Two miners died by falling down the shaft in 1889. A Finnish miner, Robert Johnson, broke his neck when he fell 35 feet from a ladder on June 4. He died the following day in the Ishpeming hospital. On October 10, Gust Dalman died when he fell 350 feet. He had only been working in the mine for two months when he lost his footing while climbing rapidly down a ladder. The 19-year-old had been warned just that morning about his unsafe conduct on the shaft ladders. The accident that killed August Trakt, a trammer, in March 1890, was caused by the typically primitive working conditions in the mine. After pushing a tram car of ore from a stope to the shaft, he was waiting for the skip when he slipped or lost his balance and fell into the unguarded shaft.

Ironically, the last three fatalities occurred during the final four months of the mine's operation. On March 23, 1897, trammer John Alio was dumping a tram car into the skip at the fifth level when the skip was hoisted unexpectedly. Alio, one of the oldest hands at the mine, was crushed between the skip and the shaft timbers. He died at the Ishpeming hospital two days later.

The final accident at the Ropes mine resulted in two deaths and may have contributed to the miners' decision to sue the company for unpaid wages later in the year. Two laborers were working in the deepest part of the mine, a 120-foot sinking stope below the 15th level. The area had been inspected by a foreman and declared safe, but about a ton of rock broke loose from the roof and fell on Henry Oja and Abram Kylmanen, killing them instantly. The two men had been working at the mine only a short time, having emigrated from Finland just two months before. The accident occurred on June 21, 1897, and the mine closed down two weeks later. In each fatality a coroner's inquest was held, with the same verdict: accidental death, the company not at fault.

After the miners' wage suit forced its closing, the mine languished in the hands of a receiver for a year. Finally, Ishpeming Postmaster Charles T. Fairbairn and attorney Albert K. Sedgwick took an option on the property. Fairbairn and Sedgwick brought the mine to the attention of several of the partners in the Michigan Copper Mining Company of Rockland, Michigan. These Detroit-based investors agreed to spend several thousand dollars to test the mine and drive a drift to a new vein which had been found west of the fifth level shortly before the mine closed. A small force of miners drifted into the vein from the fifth, seventh and 11th levels before the partners pulled out. Although the assays of the vein rock were satisfactory, there was apparently more profit to be made in the booming copper industry.

By 1899, creditors had given up hopes for the mine's reopening. One of them, the Atlantic Dynamite Company, filed suit to recover the money it was owed. Circuit Court Judge J.W. Stone ordered the Ropes company's property sold at a receiver's sale on September 5, 1899. Frederick Braastad and Sven Johnson, who held a mortgage on the property, bought the mine for $2,500. The sale was set aside by the judge, however, as the price was too low to cover the receiver's fee and the miners' and creditors' claims. A second sale on November 6 awarded the mine property to the Rand Drill Company of New York for $7,500. A month later the Rand company sold the mine to Corrigan, McKinney and Company, a Cleveland-based iron ore company, for $12,500.

Since the closing of the mine in 1897, the use of the cyanide-leaching process had revitalized the gold mining industry in the American West. Cyanide could liberate the fine particles of gold that could not be recovered by amalgamation and that were not effectively saved by

**Ropes Gold Mine 1935**
SUPERIOR VIEW

Ropes Gold Mine crew 1935
SUPERIOR VIEW

gravity concentration. Tests performed on the Ropes mine tailings indicated that there were about 100,000 tons of tailings that could yield $1 per ton profit under cyanide treatment, so Corrigan, McKinney and Company, under the direction of J. Forrest Orr and John L. Malm, erected five redwood leaching tanks, each 40 feet in diameter and 10 feet deep, along with 15 smaller intermediate and precipitation tanks. Tailings were hauled to and from the tanks in 1-ton tram cars.

Operating through the warm months of 1900 and 1901, the plant extracted a reported $54,682 in gold and silver from 30,000 tons of tailings, although in later years Leverett Ropes, son of Julius Ropes, reported that he had private information that the actual amount recovered was $194,000. William H. Rood, president of the Deer Lake Iron Company and the former manager of the Ropes mine, also set up vats to leach tailings that had washed onto the Deer Lake company's lands. Corrigan, McKinney and Company obtained an injunction to stop Rood from working, claiming ownership of the tailings. The injunction was overturned in court, however, and

81

Rood treated tailings for part of a season, recovering an undisclosed quantity of gold before his death in May 1902.

Another unexpected bonus came when John Malm inspected the old mill building, where he found discarded amalgamation plates holding a considerable amount of gold. Some of the plates had been cleaned only on the front, leaving precious amalgam on the back side. On other plates the copper had been eaten away and replaced by gold mercury amalgam. The inexperience of the Ropes company's mill men was shown by the fact that some $30,000 in gold and silver was recovered from these supposedly worthless plates when they were sent to the smelter.

Frank Lundin
SUPERIOR VIEW

Corrigan, McKinney and Company never reopened the underground workings of the Ropes mine, but finally sold the mill machinery and the 24 buildings on the property to William Trebilcock and Albert Trebilcock of Ishpeming. While tearing down the mill buildings for the lumber, the brothers found considerable amalgam under the floor, enough to bring in more money than the sale of the scrap lumber.

A pair of Ishpeming men, Frank Lundin and Albert Bjork, acquired a lease on the Ropes property in 1927. By setting up sluice boxes and siphoning water from the shaft, they were able to recover $1,000 worth of gold and silver amalgam from beneath the old stamp and vanner rooms.

In 1932, Bjork and Lundin were joined by James Trebilcock and William Trebilcock, Abel Niemi and James E. Flaa in forming the Ishpeming Gold Mining Company. Bjork, an Ishpeming businessman, served as president of the company. Lundin, a National Mine resident who had mined gold in Alaska, served as vice president and treasurer. The company reprocessed some 200 tons of tailings from an untouched portion of the tailings pile with the aid of four students from the Michigan College of Mining and Technology in Houghton. Using cyanide leaching, they recovered 9.67 ounces of gold and 14 ounces of silver, valued at $247. The same year, Bjork and Lundin leased 320 acres in Section 30, south and west of the former Phillips gold property, and explored for additional veins. On the NW ¼ of the SE ¼ of the section, they found a promising quartz vein which they traced more than 500 feet. The vein was followed by digging trenches and test pits, and a small shaft was sunk at one point, apparently without finding pay rock.

In 1934, when the government-controlled price of gold was raised from $20.67 to $35 per ounce, the Calumet and Hecla Consolidated Copper Company bought an 85 percent interest in the Ishpeming Gold Mining Company for $30,000. C & H brought in equipment and miners from their copper mining properties in Calumet, repaired the old captain's house for the miners and dewatered the mine, pumping out about 23 million gallons of water. A headframe from

Drum and disc filters
DAN FOUNTAIN

Primary grinding mill at Callahan's Humboldt mill
DAN FOUNTAIN

the Phoenix copper mine and a hoist and air compressor from the Republic iron mine were installed at the shaft. The 800-foot level was extended 300 feet to the east, and a new drift was driven 800 feet west of the shaft on the same level. Samples were taken every foot and crosscuts made every 50 feet. The ore was hauled out of the mine in blasting powder boxes and sent to the Hollinger Mine in Porcupine, Ontario, for a mill run to confirm the results of assays made by C & H.

The company also explored by diamond drill, both from the surface and from the ninth, 13th and 15th levels underground. These tests found reserves of 1.5 million tons of ore averaging 0.14 ounce of gold per ton, showing the mineralized zone to extend to a depth of at least 1,600 feet. Underground testing was completed in the summer of 1936, but the mine was kept pumped out.

In 1941, Calumet and Hecla acquired the rest of the outstanding stock in the Ishpeming Gold Mining Company and purchased the mineral rights from Republic Steel, Corrigan McKinney's successor. The company was considering reopening the mine with a new shaft and modern mill in 1942 when the War Production Board ordered all gold mines shut down to allow miners to work in the strategically important copper mines.

The property was sold by Calumet and Hecla in 1957 to two Ishpeming men, real estate agent Joe

Paul and lawyer Edmund Thomas, who resold it a few months later to Arcadian Copper Mine Tours of Ripley, Michigan. Arcadian planned to open the mine as a tourist attraction by driving a horizontal adit into the third level. Their plan was never carried out.

When the government abandoned the gold standard in 1971, the price of gold rose dramatically, eventually reaching more than $800 per ounce. These inflated gold prices made many previously marginal or unprofitable gold mines attractive to mining companies across the country. A Marquette firm, Resource Exploration Inc., initiated an evaluation program on the gold prospects in Marquette County, concentrating on the Ropes as the most favorable. They presented their findings to number of mining companies, and in January 1975 the Ropes mine was sold to Callahan Mining Corporation of Phoenix, Arizona, for about $60,000. Callahan held the property in reserve until the rising price of gold in 1979 prompted an exploration and evaluation program, initially involving geophysical and geochemical studies and the mapping of surface outcrops. The next phase

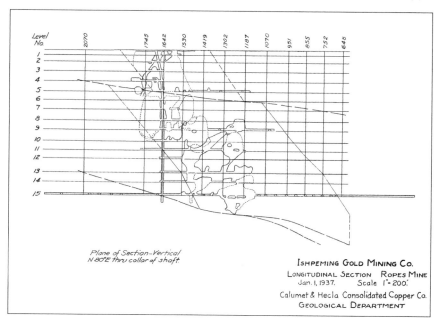

Underground workings 1937
RESOURCE EXPLORATION

of the exploration involved diamond drilling from the surface to determine the extent and richness of the gold-bearing rock.

In August 1980, pumps were installed to dewater the mine and a headframe and hoist were installed at the Curry shaft. The 400- and 800-foot levels of the mine were rehabilitated and extensive diamond drilling was carried out to detail the ore body. A bulk sample was excavated and shipped to Michigan Technological University for testing and developing an appropriate concentrating process. Total cost for these preliminary steps was $2.5 million.

In July 1983, a contract was awarded to Wallace Diamond Mining Inc. of Osborne, Idaho, to sink a spiral ramp to the 500-foot level and, on July 25, 1983, 100 years after the first run of the original Ropes mill, the sinking of the new decline began. The decline was continued to a depth of 900 feet, below the workings of the original mine. Here the first production level was developed. The ore body, which is up to 70 feet wide and extends 1,100 feet to the east, was undermined along its entire length. At the eastern end, raises were driven between the mining levels at 800 feet and above. Six-inch

Flotation cells and cyanide leaching tanks
DAN FOUNTAIN

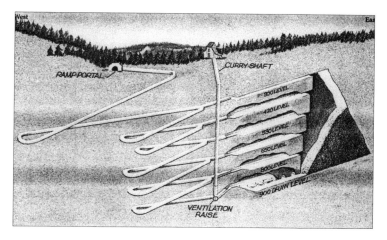

Idealized drawing of modern mining at Ropes
CALLAHAN MINING CORPORATION

blast holes were drilled and charged from the mining levels, and the ore blasted down into the undercut and raises. At first, the ore from this stope was loaded via crosscuts into 26-ton trucks in a parallel drift and hauled to the surface up the 1½-mile decline.

Although sinking a decline is an inexpensive way to open a mine, truck haulage is a very inefficient method of raising the ore. A decision was made in 1984 to excavate a vertical shaft. Rather than sinking the shaft from the surface, Callahan decided to open it by raise boring. A 12-inch pilot hole was drilled from surface to a room that had been excavated at the 1,050-foot level south of the ore zone. A gigantic reaming tool was then assembled in the room and slowly hoisted toward the surface, drilling out the rock as it went and leaving a 14-foot diameter, untimbered shaft. A used headframe was acquired from the Lakeshore Copper Mine near Casa Grande, Arizona. This headframe had originally been manufactured by Lakeshore Inc. of Kingsford, Michigan. The old but reliable hoist purchased for the Ropes shaft had been built in the 1930s by Nordberg of Milwaukee and was one of the first generation of electric hoists built by the company. Although the shaft and associated equipment cost $3.8 million, it was estimated that a $35-per-ounce saving could be realized by hoisting rather than trucking the ore to the surface. Equipment was still driven in and

Ropes Gold Mine
1955

out of the mine via the decline, but the ore was hoisted up the shaft in 15-ton capacity skips.

As the ore above the 900-foot level was depleted, the decline was extended, with draw levels being developed at the 1,284- and 1,548-foot levels. Ore was mined from a stope extending up to the 1,020-foot level much as it was from the upper levels in the mine. The decline was continued still deeper, heading toward an ore body beginning at a depth of about 2,000 feet. Diamond drilling has shown the ore in this zone to be comparable in width and richness to that in the upper levels.

From the mine, the ore was trucked some 15 miles to Humboldt, where Callahan had purchased and renovated the former Humboldt Mining Company iron ore concentrator. Part of the trip was made over a new road constructed on an old railroad grade between North

Lake and the Verde Antique marble quarry just west of the gold mine.

Utilization of modern milling and recovery techniques allowed Callahan to make a profit treating ore that had been uneconomical for the original operators of the mine. While the amalgamation and gravity concentration methods used by the miners of the past century could only recover the coarser gold from the quartz vein rock, the modern chemical process could treat all of the rock, vein quartz and greenstone country rock alike.

At the Humboldt plant, the coarse ore from the mine was crushed in two stages and fed to two 10½-foot-diameter by 16-foot-long ball mills for fine grinding. A ball mill uses a rotating drum partially filled with hard steel balls to grind the ore. Xanthate, a flotation reagent, was added to the finely ground ore, and air was bubbled through the mixture. The gold, along with its associated pyrite and copper minerals, would cling to the air bubbles and float to the top, leaving the waste rock behind. After more gold-bearing minerals were floated off in two additional flotation stages, the waste rock tailings were piped to the flooded Humboldt mine pit for disposal. The gold-bearing flotation concentrate was ground to a finer consistency in a 10-foot diameter, 6-foot-long regrind ball mill, then dewatered in a bank of vacuum disc filters. Next, it was mixed with a dilute solution of sodium cyanide and aerated for 24 to 48 hours until the gold and silver were leached out of the ore. The gold-bearing cyanide solution was separated from the concentrate in vacuum drum filters. The concentrate was then mixed with fresh cyanide solution and the leaching and filtering process was repeated to recover additional gold and silver values.

After the gold-bearing cyanide solution was clarified and deaerated, powdered zinc was added, causing the gold to precipitate out of solution. This precipitate, which contained gold and silver along with traces of other metals, was finally melted down and cast into buttons of doré bullion. The bullion was sold to refineries for further purification. Once the gold was leached out

Geologist's office perched on the edge of the cave-in
BOB PROCUNIER

of the concentrate, copper, silver and pyrite were left in the concentrate. These minerals were separated out in two further stages of flotation and dried. The resulting copper/silver/pyrite concentrate was sold to the Copper Range Mining Company's White Pine copper smelter, where it was a valuable addition to the smelter feed.

Mining today is generally much safer than it was a hundred years ago, due to federal mining laws, company safety policies and the use of machinery in place of manpower. Despite these improvements, underground mining continues to be a hazardous occupation, and the modern Ropes mine had its share of accidents. On June 4, 1987, a miner was operating a jumbo drill rig in the haulage level 1,284 feet below surface. Suddenly, a large mass of rock let go from the roof and fell on the jumbo, partially burying it and caving in part of the protective canopy over the operator's compartment. Fortunately, the operator escaped injury. Nine days later, however, another miner was not so lucky. While he and his partner were installing rock bolts in an attempt to secure some unstable ground in the 1,154-foot level, some rock broke loose and fell. The falling rock knocked the miner down and buried him to his waist, breaking his left leg in several places. Another miner was injured by falling rock on August 29, 1987. The miner was installing roof bolts in an unstable area on the 420-foot level when some loose rock fell, breaking bones in both of his feet.

The most dramatic accident at the Ropes mine occurred on the morning of December 31, 1987. Shortly after the day shift arrived at work, surface workers heard a rumbling noise, dust began to spew from the Curry shaft, and the ground near the shaft began to cave in. Although the production stope of the modern mine began at a depth of 300 feet and went down to 900 feet, the first level of the old mine was only 30 feet below the surface. The original miners had mined out the upper levels of the mine, leaving great open stopes reaching from the first level to at least the fifth level at 250 feet. As the ore body in the modern stope was mined out up to the 300-foot level, the pillar of rock between the old and new stopes was left unsupported. When it finally collapsed into the stope, the old workings, honeycombed with drifts, crosscuts and stopes, caved in with it.

As the employees watched the gaping hole grow, a delivery man bringing supplies to the mine began to drive on the entrance road toward the shaft house and the mine buildings beyond. Mine workers stopped the driver and told him to back up. The driver parked his van and was waiting for permission to enter when the ground gave way beneath the truck, plunging him into the collapsing mine. Stunned and bleeding, he struggled out of his truck and scrambled to safety up the side of the still-crumbling cave-in. Fortunately, his injuries were minor, requiring only a few stitches. Nobody else was injured, but a power line, the mine geologist's office and the Curry shaft house were swallowed up in the 200-by-700-foot-long, 100-foot-deep hole.

The cave-in did not damage Callahan's production shaft, which is located about 200 feet south of the stope. The shaft was rendered unusable for several days, though, since the power line supplying the hoist was lost in the accident. Once power was restored, company officials and federal mine inspectors were able to inspect the mine. They found no damage to the working areas of the mine, but, due to safety considerations, it was about two months before full production could be resumed.

In order to stabilize the ground surrounding the caved stope, Callahan had 350,000 cubic yards of sand trucked in and dumped into the hole. Until the stope was filled, no ore could be mined from the production levels at 900 feet and below. Low-grade ore from stockpiles at the mine was processed at the plant, along with higher-grade ore from a small ore body northwest of the main workings. This ore was mined from a crosscut from the decline and trucked to the surface. To provide safe access to the levels below the 900-foot production level, an alternate leg was added to the decline, running from the 900- down to the 1,020-foot levels.

Although the cave-in was an expensive accident for Callahan, it did have one potential benefit. The original plan for the mine called for ore to be pulled from the bottom of the stope as it was blasted loose, leaving a great empty hole along the ore zone. Early on, it was discovered that the wall rock surrounding the gold ore was not strong enough to stand on its own. The sides of the stope began to cave in, diluting the gold ore with worthless rock. To support the stope walls, production was halted for a month in the spring of 1986 while ore in the upper levels was blasted loose to fill the stope. To avoid similar problems in the lower levels, a pillar of ore was to be left between the 900-foot production level and next stope being developed from 1,020 to 1,548 feet. Once the upper stope caved in, it became possible to change the mining system to one of continuous caving, backfilling the stope from the surface as ore was pulled from the bottom. With the entire stope stabilized with fill, the pillar of ore between 900 and 1,020 feet could be blasted loose and recovered.

In October 1989, Callahan announced the indefinite shutdown of the Ropes mine. An unstable zone of rock where the gold-bearing formation meets the surrounding

**Headframe 1982**
DAN FOUNTAIN

country rock was threatening to allow a section of the untimbered shaft to collapse. The low grade of ore the mine had been producing and the price of gold were undoubtedly also factors in the closing. Original projections for the mine were that the ore would average 1 ounce of gold per 10 tons of ore. When it was shut down, however, it had been yielding only 1 ounce for every 17 tons milled. The price of gold, which was at $324 per ounce when the mine opened, had risen to $450 by early 1987, the mine's most profitable year. The price had slid to $366 when the shutdown was announced.

    Hoisting was stopped in October to allow the company to reinforce the shaft. Within the unstable zone, 300 feet of the shaft were to be lined with steel and grouted with concrete. Bids for the reinforcement work came in too high, however, and Callahan decided on a contingency

Underground loader entering the decline portal
DAN FOUNTAIN

plan to allow operation of the mine if economic conditions improved. Rather than reinforce the unstable portion of the shaft, which was in the deeper reaches of the mine, the company decided to abandon the lower portion of the shaft and to truck the ore part way up the decline to the original skip loading pocket at the 900-foot level. To provide the necessary alternate escape route formerly provided by the shaft, raises were excavated between the decline legs in the lower levels of the mine.

Gold prices failed to rise sufficiently, however, and in April 1990 Callahan Mining Corporation announced that the Ropes mine and mill were being mothballed. In the spring of 1991, mining equipment, including dewatering pumps and the underground primary crusher, were removed, and the mine was left to slowly fill with water. The mining equipment, haulage trucks and mine hoist, as well as the Humboldt processing plant, were put up for sale. The waste rock dump was covered with soil and revegetated, and the shaft and decline portal were sealed, ending another chapter in the gold mining history of Michigan.

## Ropes Gold & Silver Company Production 1883-1897

| YEAR | TONS | GOLD | SILVER | PRODUCTION | YIELD/TON |
|---|---|---|---|---|---|
| 1883 | 272 | $1,651.16 | $358.81 | $2,009.97 | $7.39 |
| 1884 | 1,371 | $6,163.19 | $616.20 | $6,779.39 | $4.94 |
| 1885 | 5,643 | $23,288.92 | $2,942.19 | $26,231.11 | $4.65 |
| 1886 | 6,959 | $38,499.93 | $4,653.92 | $43,153.85 | $6.20 |
| 1887 | 10,216 | $32,338.63 | $2,592.03 | $34,930.66 | $3.42 |
| 1888 | 16,855 | $52,353.94 | $5,330.81 | $57,684.75 | $3.42 |
| 1889 | 31,365 | $90,060.40 | $9,655.05 | $99,715.45 | $3.18 |
| 1890 | 31,578 | $71,132.70 | $8,435.57 | $79,568.27 | $2.52 |
| 1891 | 21,355 | $63,760.59 | $8,025.39 | $71,785.98 | $3.36 |
| 1892 | 21,794 | $52,443.08 | $7,160.82 | $59,603.90 | $2.73 |
| 1893 | 15,080 | $40,416.56 | $3,509.21 | $43,925.77 | $2.91 |
| 1894 | 21,185 | $43,676.77 | $1,789.00 | $45,465.77 | $2.15 |
| 1895 | 16,535* | $34,838.69 | $1,373.16 | $36,211.85 | $2.19 |
| 1896 | 16,686 | $37,235.98* | $1,489.44* | $38,845.42 | $2.32 |
| **Total** | 216,894 | $587,860.54 | $57,931.60 | $645,792.14 | $2.98 |

*Estimated.
1897 Figures not available.

# CHAPTER 5

## The Michigan Gold Mine

**Anson B. Miner**
ALL IMAGES AUTHOR'S
COLLECTION, UNLESS
INDICATED

Anson B. Miner, the cashier at the Ishpeming National Bank, had lived in the West for several years and was familiar with gold prospecting. With the Ropes Gold Mine coming into its own in 1885, Miner and several companions spent their spare time searching for other gold veins. While prospecting some three miles west of the Ropes, an outcrop of rusty looking, yellow-stained quartz caught his eye. On Sunday June 14, 1885, Miner and his partners set off a blast in this vein, which broke out a large quantity of quartz with glittering particles of gold protruding from it. Miner's group tried to get a lease on the land from its owner, the Lake Superior Iron Company, but a company policy forbade the granting of leases for mineral exploration. Unable to pursue their find on the iron company's land, the prospectors traced the vein a quarter mile to the east onto private land on the NW ¼ of the NE ¼ of Section 35, T48N-R28W.

The landowner, Peter Gingrass, who ran Ishpeming's Urban House hotel, was more than happy to grant the prospectors a lease on his forty – for a price. The Miner partnership, which included bank director Edward R. Hall, iron mining agent James Jopling, assayer Will L. Jones and W.L. Lyons, immediately sold an interest in the property to Peter White, pioneer businessman of Marquette, and started sinking a shaft. About this time it became known that Gingrass had granted an earlier option to Joseph C. Foley, but had subsequently refused him a lease, due to his lack of progress in exploring the land.

Interest in the property lagged for two years until gold-bearing quartz worth $44,000 per ton was discovered in the adjacent Lake Superior Iron Company's shaft in July 1887. Within a week, Foley set a crew of men to work crosscutting the vein, but Miner, accompanied by Peter White's lawyers, evicted Foley's men and set up a camp of his own, a move characterized in a local newspaper as "claim jumping."

By this time, a group of Cleveland iron and steel industrialists, headed by Colonel James Pickands and Samuel Mather of the powerful iron-ore mining company Pickands, Mather and Company, had bought the Miner and White lease and formed the Michigan Gold Company. The company renegotiated the lease with Gingrass and extended it to 40 years for an immediate payment of $10,000 and a 10 percent royalty. Frank P. Mills, the superintendent of the Cleveland iron mine, was put in charge of the Michigan Gold Company's property.

A third claimant, George Grummett of Ishpeming, entered the fray when he had the five men working

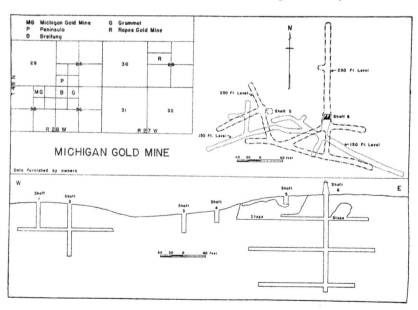

Michigan Gold Mine plans and section.

**J. C. FOLEY,**

**FIRE INSURANCE and REAL ESTATE AGENCY**

MINERAL AND PINE LANDS FOR SALE.

Loans Negotiated.   Rents collected.   Mining Stocks bought and sold on commission.
Patronage solicited with a guarantee of prompt attention.

**Office, Voelkers Block,   -   -   ISHPEMING, MICH.**

for the Michigan company, as well as Foley's two men, arrested for trespassing. Grummett, who was backed by Captain William Ward of Pittsburgh, claimed to have a lease on the property from Gingrass, but its validity was questionable. The arrest of the Michigan and Foley parties was thrown out by Justice of the Peace Leonard P. Crary of Marquette, but Grummett was able to make a second arrest stick, with the men being fined $10 each by Justice of the Peace Stearnes of Clarksburg. Not coincidentally, the settlement and blast furnace at Clarksburg were owned in part by the Ward family. In October, the powerful Michigan company had Grummett and his superintendent, William Gardiner,

# Urban ✢ House

**P. E. GINGRASS, Proprietor.**

Rebuilt, Thoroughly Renovated and Newly Furnished Throughout.

Located in the Business Centre of the City, One Block from the South Shore & C. & N. W. Depots.

**RATES · $2.00 · PER · DAY**

FIRST-CLASS SAMPLE ROOMS ON FIRST FLOOR.

Special Rates by the Week.            **ISHPEMING.**

arrested and evicted, effectively gaining control of the property. The Michigan company built a boarding house and sank two shafts, the No. 1 to 76 feet and the No. 2 to 40 feet, before suspending operations for the winter in January.

The spring of 1888 found the Michigan company reopening and dewatering its shafts on the gold range, and defending itself against a lawsuit by Grummett in the courts at Marquette. Work on the property was limited to exploration and prospecting pending the outcome of the lawsuit. In August, Judicial Court Judge C.B. Grant delivered a decision in favor of the Michigan Gold Company, but work was still held to a minimum because of an appeal to the state Supreme Court by Grummett and Ward. Grummett, meanwhile, was sinking a shaft on another vein half a mile to the east.

Peter White

The truly rich nature of the Michigan mine became apparent in late August when an exceptionally rich lode of quartz was found near the surface in the No. 2 shaft. Pieces of rock the size of walnuts were found to carry up to 1.5 ounces of gold, and pieces of shattered quartz were held together by the soft, flexible native metal. Assays of the rock showed as much as $100,000 worth of gold per ton, a phenomenal amount for any gold mine. This rich lode was followed to a depth of 50 feet by the end of 1888.

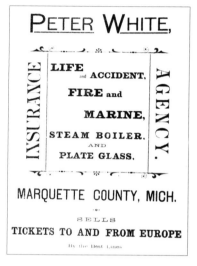

In January 1889, 70 tons of rock from the Michigan mine was run through the stamp mills of the Ropes Gold and Silver Company as a test of the average value of the deposit. The ore proved to be easily crushed and yielded $6.68 per ton, still a respectable grade of ore. Rich specimens continued to be found in several shafts as the Michigan company awaited to hear from the court. Finally, on November 8, 1889, the Michigan Supreme Court handed down a decision in favor of the Michigan company, which within two weeks installed boilers, steam engines and a hoisting plant. Mining began in earnest.

In a demonstration of the mine's potential, about 200 pounds of rock were brought to Ishpeming, coarsely crushed and washed in a hand rocker. Even by these crude concentrating methods, 137 ounces of gold were extracted, a yield of $30,000 per ton. The gold was cast into two bricks, each 8-by-2-by-½ inch, and put on public display. With this proof and the company's clear title to the mine, Michigan Gold Company stock skyrocketed, increasing 500 percent in one 10-day period.

A force of 18 men went to work, continuing the sinking of the rich No. 2 shaft. They also erected a mill building housing a Blake jaw crusher and secondary roll crusher, which fed a 3½-foot Huntington centrifugal grinding mill. The company's intention was to mill the rich rock as it was mined to cover expenses until the extent of the gold deposit was known and a permanent stamp mill of an appropriate size could be erected. However, the mill and its associated equipment didn't run as smoothly as expected since the mill was too small. As a result, the company had to call in a 10-cent-per-share assessment in May 1890 to cover expenses. Milling the rock from shaft sinking and drifting yielded $12 to $24 per ton.

The mining company also underwent a change of management in May, the Cleveland directors giving way to Upper Peninsula influence. David M. Ford, a prosperous stockbroker, and James R. Cooper, the superintendent of the Calumet and Hecla copper smelter, both of Houghton, were elected to the board. They were joined by Peter White of Marquette,

Captain Sam Mitchell

James Cooper

the famous iron-mining captain Samuel L. Mitchell of Negaunee, and Ishpeming merchant A.W. Myers.

The Michigan mine's managers had long suspected that some of the miners were "high grading," that is, stealing rich specimens of gold. Miners would work for a few months, secretly carrying out hundreds of dollars worth of gold in their lunch pails, then suddenly quit and leave town. In May 1890, a pair of such "high graders" were arrested on the eve of their departure for England. Henry Varcoe, who had been a shift boss at the mine, and his son William were caught with more than 200 pounds of rich quartz in their steamer trunks, estimated to be worth anywhere from $1,000 to $5,000. The next day, William Uren, who ran the boardinghouse at the mine, was arrested when several ounces of coarse gold were found in his possession. The Michigan Gold Company was unable to prove in court that either the Varcoes' or Uren's gold had come from the Michigan mine, however, and all three were acquitted. The mining company instituted tighter controls on its miners and erected a barbed-wire fence surrounding the vein for its entire length.

Rich specimens notwithstanding, the total supply of

Michigan Gold mine 1935
MARQUETTE COUNTY HISTORICAL SOCIETY

gold at the Michigan mine remained in doubt, prompting the company to make another test run of low-grade rock at the Ropes mill in September 1890. The run yielded only $2.82 per ton, indicating that the high-grade rock had pinched out. Another 10-cent-per-share assessment was called to provide capital to install an air compressor and power drills and to sink the shaft farther in search of more rich deposits. In February 1891, a third mill test of 1,000 tons was ordered at the Ropes. However, the yield from the rock tested was so low – 71 cents per ton – that the test was stopped after only half that amount had been stamped.

The Michigan Gold Company then hired the Ropes company's manager William H. Rood and superintendent George Weatherston to systematically sample and test various veins and openings on the property. After nearly a month, Rood and Weatherston submitted their report, having found an average of less than $1 per ton in shafts 1, 2 and 3. Only the vein at the No. 4 shaft and a small shaft nearby was considered worth mining. Accordingly, the company moved their portable hoist to the No. 4 shaft, dewatered it and continued sinking. A rich seam of pay rock produced $1,000 per month for a few months, enough to cover expenses.

During 1892, 15 cents per share in assessments was called to provide operating capital for a reduced force of miners. It was decided that a surface crew of seven men to support only four miners was too expensive, since it cost the company $53 per foot of drifting. The company took bids for a contract for hand-powered drifting, estimating that a two-thirds saving could be realized.

Although another streak of rich ore was found near the No. 3 shaft and a new small vein was found to the south, the mine was shut down for the winter in January 1893. The mine wasn't reopened in the spring as planned, however. Although the workings were pumped out in late June, unfavorable economic conditions brought on by the Panic of 1893 prompted company president Peter White to decide against the resumption of mining. The management still had faith in the property, though, since the company renewed its option that summer and talked of putting in a five-stamp mill.

The mine was still idle in the summer of 1894 when Thomas Trevithic, the former mining captain, brought specimens of quartz with native gold assaying $150,000 per ton to the company's attention, claiming to have found a new vein on the Michigan property. He offered to reveal the location of the vein for a $250 fee and the promise of six months work, but, distrusting the company (and Peter White in particular), later raised his price to $1,000. The company chose not to pay him, but instead set several of their own men to work prospecting for more rich veins. Enough pay rock was found to keep the mine working intermittently until January 1896, when the Michigan Gold Company closed down for good. Total reported production was $17,699.36, although an unknown amount in rich specimens had also been stolen from the mine and ore piles.

In 1901, Peter Gingrass, the owner of the Michigan property, approached several groups of investors in the

Copper Country in an attempt to reopen the mine. He gave an option to Will A. Bateman of Calumet and L.C. Fredericks of Colorado, but no work was done. Another attempt to reopen the mine was made in 1902 by Gingrass' son-in-law, Edward Copps, who had been an officer of the Wakefield Gold and Silver Company during the Wakefield gold and silver mining boom in 1887. Copps tried unsuccessfully to raise $25,000 to incorporate the Pere Marquette Mining Company to work the Michigan mine. The next year, however, he did persuade the Tribullion Mining, Smelting, and Development Company, which had mining properties in Arizona and Montana, to lease the property and conduct explorations.

The Tribullion company dewatered the old No. 1 and 2 shafts, finding rich quartz in both shafts and in the drift between them. The drift was extended to the east over the next few months until, in December 1903, the air compressor supplying the power drills broke down. Even though they were finding rich quartz, the company decided to allow its lease to lapse in order to concentrate its efforts on the copper mines in the West.

Although the rich pockets of gold were elusive, the abundant white quartz had economic value of its own. When finely ground and screened, it sold for $18 to $40 per ton, for use as a filler and abrasive. Seeing the possibility of mining quartz for sale as silica while still seeking rich pockets of gold, Edward Copps and his associates formed the Ishpeming Gold Mines Company in 1905. Copps served as president of the new company, with E.D. Nelson, a former Ishpeming resident who had become a prominent banker in Iron River, as vice president, Peter Gingrass's son Joseph as the secretary, and William

---

### THE
# Michigan Gold Co

### VICTORIOUS !

In the SUPREME COURT, whose decision last Friday wipes out both the

**Grummet and Foley Claims.**

---

Mining work in full blast. A new Stamp Mill to be built. GOLD enough on hand to pay for it. Three more chimneys of

## RICH GOLD FOUND.

---

NOW is the time to buy stock in this BONANZA MINE before it goes higher. If you want to buy, or have stock for sale, address  DAVID M. FORD,
Houghton, L. S., Mich.

Miners at Michigan Gold Mine 1935
ETHEL VALENZIO

Noon of Marquette as treasurer. Other directors included Peter White, Otto Eger of Ishpeming, and George Russell, a banker from Detroit. The company bought the Michigan, Peninsular, Superior and Grummett gold properties from Gingrass for $25,000 and began development work at the Michigan mine.

Over the next few years, the Ishpeming Gold Mines company reopened two of the shafts, built a mill building and installed the equipment necessary to grind the quartz and separate any precious metal or impurities before shipping the silica powder. The mine and mill finally went into operation in the summer of 1908.

Quartz was mined from the No. 2 shaft and taken from the dumps left from earlier operations. After crushing, the ore was fed to a Huntington centrifugal mill, which ground the quartz and separated some of the coarse gold. More gold was separated from the ore on vibrating tables, and iron and other heavy minerals were removed in a Frue vanner. The pure quartz was then finely ground in a tube mill, dried, and barreled for shipment.

The mine was sold in 1909 to the Michigan Quartz Silica Company. This Milwaukee-based company, of which Copps was also an officer, operated the property intermittently for the next few years. The Milwaukee company eventually shut down the Michigan operation, finding it less expensive to buy quartz from Wisconsin and Illinois and treat it in its mill in Milwaukee.

The mine did not attract any further attention until 1933, when a group of investors formed Michigan Gold Mines Inc. These investors, led by Henry B. King of Three Rivers, president; Wesley B. Orr of Manistique, vice president; Frank Trombley of Marquette,

LEE DEGOOD

treasurer; and Cora Secor of Gladwin, secretary, leased the Michigan mine and shipped 10 tons of rock from the mine dump to the Michigan College of Mining and Technology for testing.

The new company erected a headframe at the No. 6 shaft and started sinking below its original depth of 53 feet. Within a few feet, the "pinched out" vein opened out to 5 feet wide. The shaft was deepened to 270 feet, and 1,000 feet of drifting and development work was carried out on the 150- and 250-foot levels. A mill building was also erected, in which the ore was processed by first crushing it in a jaw crusher, then finely grinding it in a 72-by-36-inch Hardinge ball mill. The gold was separated from the rock by gravity concentration on corduroy blankets, then amalgamated with mercury. The first shipment of bullion, consisting of a brick weighing 18 ounces, was made in April 1934. The value of the gold produced was enhanced by the 1934 increase in the statutory price from $20.67 to $35 per ounce.

During the summer of 1934, the Michigan College of Mining and Technology held a mill practice course at the Michigan mine, experimenting with methods for extracting gold from the ore. Four shipments of bullion were made to the U.S. Mint in New York that year, totaling 58.63 ounces. In May 1935, the company was reorganized as the Michigan Gold Mining Company. Henry B. King stayed on as the president and general manager and Samuel M. Cohodas of Ishpeming joined as vice president. The rest of the directors came from Milwaukee and the Appleton area of Wisconsin, including A.F. Schroeder, treasurer; Clarence F. Manser, secretary; Walter L. Schroeder; and Dr. J.G. LaHam. Kiril Spiroff of the Michigan College of Mining and Technology in Houghton served as consulting engineer for the company.

No mining or milling was done in 1935 and 1936, but exploration and development work was carried out underground and a 100-ton flotation plant installed on the surface. The mill ran for a few months in 1937, but only produced 51 ounces of gold, worth about $1,800. The company then shut down, leaving the miners unpaid

Miners at Michigan Gold Mine 1935
ETHEL VALENZIO

Michigan Gold Mine
1935
SUPERIOR VIEW

when one of the company officials reportedly absconded with all the gold recovered from the mill in the final cleanup. The following year, a bolt of lightning struck the power lines leading to the mine, burning out the electric motors in the mill. The company was reorganized once again late in 1939, this time as Marquette Mines Inc. The new company paid off the mine's outstanding debts, but never reopened the property.

The renewed interest in the Ishpeming Gold Range in the 1970s and '80s extended to the Michigan mine, which was included in the preliminary studies leading up to the purchase of the Ropes Gold Mine by Callahan Mining Corporation. Callahan purchased the Michigan property in 1985 but didn't find the property promising enough to consider reopening it. The mineral rights were bought by Minerals Processing Corporation along with much of Callahan's other mineral lands on the Michigan Gold Range. As of 2013, the property was under option to Aquila Resources Inc., but no active exploration was being done.

# CHAPTER 6

# Michigan Range Prospects

Although the Ropes mine proved that gold could be found in Michigan, it was the rich finds at the Lake Superior gold mine that brought national attention to the Ishpeming Gold Range. In July 1885, prospectors discovered gold not far from Deer Lake on the NE ¼ of the NW ¼ of Section 35, T48N-R28W, owned by the Lake Superior Iron Company. Since the mining company would not grant mineral leases, the explorers traced the vein onto land owned by a private individual, where the famous Michigan Gold Mine was eventually started.

The Lake Superior Iron Company owned and operated several large iron ore mines in Ishpeming and was less than enthusiastic about gold mining, especially since company President James S. Fay of Boston had lost money in an ill-advised gold venture in Colorado. The company planned to test the vein by diamond drilling, but its drill rigs were in constant use exploring for iron ore. Finally in June 1887, an exploratory shaft was started. It had been sunk 22 feet on the vein when an 8-inch rich streak was found, holding up to $44,000 per ton in gold and silver. The rock was so filled with gold that large chunks of shattered quartz were held together by the soft, flexible metal.

News of the rich find traveled quickly, broadcast as far away as New York. The Lake Superior company temporarily closed the shaft, posted a guard and built a stockade around the stockpile. Organized as a stock company to mine iron ore, the Lake Superior Iron Company had to reorganize its charter to go into the gold mining business.

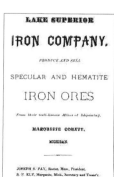

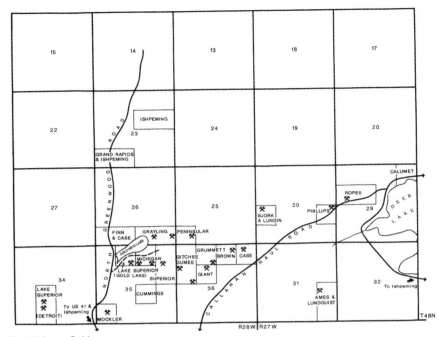

The Michigan Gold Range
NMU CARTOGRAPHY LAB

Once their legal position was secured, the company continued to explore for gold, sinking a 25-foot shaft on a second vein 100 feet to the south. In the spring of 1888, the original shaft was dewatered and enlarged under the supervision of Captain John Jenkins, and a steam hoisting plant was installed. More rich rock was found at 35 feet, assaying as high as $60,000 per ton. The shaft was deepened to 100 feet over the next year, and drifts driven at the 40- and 100-foot levels, but no more rich streaks were found. The quartz in the lower reaches of the mine still averaged $6 per ton, however. The majority of the company's stockholders remained opposed to the relatively speculative business of gold mining, and at the June 19, 1889, stockholders meeting in Boston, voted to sell or lease the gold prospect to outside developers. Operations at the shaft ceased two months later.

Four months after the Lake Superior Iron Company shut down operations, the Gold Lake Mining Company was formed to operate the mine. Officers included George

W. Hayden of Ishpeming, president; N.H. Stewart of Kalamazoo, vice president; Charles T. Fairbairn of Ishpeming, secretary; W.C. Dewey of Grand Rapids and Frederick Braastad of Ishpeming, directors. The company was capitalized at $2.5 million and leased the Lake Superior forty and the adjacent one to the south.

After the shaft was dewatered, the upper drift was extended 40 feet to the east, striking a rich chimney of ore. The shaft was deepened to 110 feet and the lower drift was extended, following the vein to the west. Three hundred tons of ore were stockpiled, awaiting the purchase of a mill. In March 1890, a test of 50 tons of quartz was made in the Michigan Gold Company's mill. From ore showing no coarse gold, an average of $15.25 per ton in bullion and $5 per ton in concentrate was recovered. The April snow melt flooded the lower levels of the mine, however, and the last mention of the mine in local newspapers indicated that the company planned to strip the vein on the surface. The Gold Lake Mining Company surrendered its lease in 1898, and the mine was never reopened, remaining a potential bonanza to this day.

Many of the early gold mining companies failed long before their properties had been thoroughly explored or developed. The evolution from a promising prospect to a paying mine is an expensive proposition. Even the Ishpeming Gold Range's most successful mine, the Ropes, eventually failed, in part because the company lacked the money needed to develop the mine for efficient operation.

A strong capital foundation, however, was no guarantee of success, as demonstrated by the fortunes of the Superior Gold and Silver Company. Although vigorously backed by such wealthy investors as Marquette banker Peter White, and owning promising prospects in

Frederick Braastad

ALL IMAGES AUTHOR'S COLLECTION, UNLESS INDICATED

Nathan M. Kaufman

three different areas, the company never opened a successful gold mine. Organized in late August 1888, the company leased the NE ¼ of the NE ¼ of Section 35, T48N-R28W, immediately east of the Michigan Gold Company's mine. The quartz vein in which the Michigan company had found its rich free gold had been traced onto this 40-acre tract, where it was hoped similarly rich deposits would be found. In fact, ore averaging up to $20 per ton was found in a 30-foot shaft on the main vein and on several smaller veins in 1888 and early 1889. Another vein, hoped to be a continuation of the Peninsular formation, was found in the northeast corner of the property. None of these veins, however, proved to be large enough to pay for mining, being either too narrow or too inconsistent. In 1890, the Superior company obtained an option on the next forty to the south from the Lake Superior Iron Company. What was believed to be an extension of the Michigan vein on this parcel was stripped and sampled, but this vein, too, failed to yield the anticipated riches.

The Superior Gold and Silver Company later explored prospects on the Ropes and Dead River Gold ranges, but these ventures also met with failure. The company, organized by Peter White, Edward R. Hall, Clarence R. Ely and Anson B. Miner, was capitalized at $2.5 million. Marquette businessman Peter White was the major stockholder. Officers included Dr. J.V. Vandeventer as president; C.R. Ely as secretary: and A.B. Miner as treasurer. The Superior property was acquired by the Michigan Quartz Silica Company in 1909, and in 1916, the company attempted unsuccessfully to extract gold from Superior ore in their mill at the Michigan mine. Callahan Mining Corporation purchased the Superior along with the Michigan mine in 1985, and it was the site of a program of diamond drilling during 1987 and 1988. The new owners hoped they would find the rich veins that

had eluded the Superior company a century before, but no work was done beyond exploratory drilling.

East of the Superior Gold and Silver Company's property on the Michigan range was a prospect worked by the Gitchie Gumie Gold Mining Company. The W ½ of the NW ¼ of Section 36, T48N-R28W had been explored in 1885 by R.D. Vaughn and his associates. Assays of the quartz vein found there reportedly showed enough gold to pay for mining. Vaughn apparently let his option lapse, however, since the next mention of the property, which was owned by Edward Breitung, a pioneer Negaunee mining man, had the option held by Ephraim Coon. Coon sold his option to Robert Nelson and Judge Henry H. Mildon in August 1887. A year later, Mildon, along with Coon, W.P. Healy, E.P. Bennett and Nathan M. Kaufman, formed the Gitchie Gumie Gold Mining Company and exercised their option to take a 20-year lease on the northern half of the parcel. Here they erected a blacksmith shop, supply house, and a boarding house for the miners. The vein, which was up to 6 feet wide, was stripped of overburden for 500 feet, and the quartz was mined from the most promising areas of the vein. A 40-foot shaft was sunk at one point. Apparently

George Grummett
MARGUERITE GRUMMETT BERGDAHL

the vein did not prove rich enough to pay, since by July 1889 all work had stopped.

While George Grummett contested the rights to the rich Michigan vein in the courts, he was also exploring the NE ¼ of the NW ¼ of Section 36, T48N-R27W, a half mile east of the Michigan mine. In December 1886, Grummett had obtained an option to explore the forty from its owners, Peter and Victoria Gingrass, and leased the parcel a month later. Over the next two years, Grummett opened several test pits and three shafts, the deepest of which eventually reached 62 feet. Some high-grade specimens carrying free gold were found, but the vein as a whole was not rich enough to warrant expansion and by 1889 the prospect had been abandoned.

After discovering the Lake Superior and Michigan gold veins, A.B. Miner and E.R. Hall continued to explore on the Michigan range. On the SE ¼ of the NW ¼ of Section 36, T48N-R28W, they found a four-foot-wide vein of gold-bearing quartz, apparently a continuation of the Grummett vein. Here they sank a shallow shaft. Late in 1888, they sold the prospect to a group of Chicago and Upper Michigan investors, who formed the Giant Gold and Silver Mining Company. George and Albert Raymond of Chicago had invented a machine to separate gold from its ores, and they saw the prospect as a way to prove their machine. Along with their partners, George Parmlee of Chicago, H.J. Payne of Escanaba, John McDonald of Iron River, and James H. Molloy of Ishpeming, they set six men to work under Captain Richard Trevarthen. No free gold was found in the shaft, although the one assay report made public showed values of $10.40 per ton. The prospect was abandoned in 1889, and the Raymond brothers had to look elsewhere to demonstrate their patented gold separator.

A mile and a half east of the Michigan mine, James and Thomas Dwyer, wholesale liquor dealers from Marquette, found a mineralized quartz vein on their land in 1885. The vein, on the NE ¼ of the NE ¼ of Section 36, T48N-R28W, produced surface samples which assayed up to $11 per ton. Captain William Ward bought a quarter interest in the property and financed further prospecting. A test pit was sunk about five feet into the vein before exploration was halted. Hosea B. Swain, a professional mineral prospector, explored the property for a syndicate of Marquette men in 1888. Working with his brother, Swain traced the vein for 400 feet on its course across the forty. Several trenches and shallow test pits uncovered gold-bearing quartz assaying up to $121 per ton. Swain's backers refused to put up any more money, however, so he sold his option to Julian M. Case of Marquette.

Case bought Ward's interest in the property and immediately set his men to work, sinking a shaft and building a blacksmith shop, supply house, and a boarding house for the miners. The shaft followed the vein to a depth of more than 30 feet and produced some finely disseminated free gold. The gold-bearing quartz was mixed with slate, however, and eventually pinched out at depth.

Francis F. Palms

Between the Grummett prospect on the west and the Case on the east was a forty known as the Brown prospect. About 1888, unknown parties found several promising quartz veins and sank some pits and a shallow shaft. The vein was found to be narrow and inconsistent, and work was soon abandoned. It is not known for whom the prospect was named.

One of the most promising gold prospects on the Michigan range was the Peninsular mine. Discovered in 1885 by John Sanson, this prospect was located in a large reef of quartz which was up to 50 feet wide. An experienced prospector, Sanson had been exploring west of the Ropes mine for several months on lands owned by Peter Gingrass and the Breitung estate when he discovered the gold-bearing quartz in the SW ¼ of the SW ¼ of Section 25, T48N-R28W. He continued to explore the area for two years, selling a less promising prospect to help raise enough capital to purchase an option from Gingrass and to start mining. Working with four men, he drove an adit a short distance

Diamond-drill work at the Peninsular site
DAN FOUNTAIN

into the side of the bluff and sank a shaft 20 feet, finding ore averaging $20 per ton before his money ran out.

Sanson sold the prospect in September 1888, to Alfred Magoon of DePere, Wisconsin, who in turn sold it to a group of Detroit investors who incorporated the Peninsular Gold Mining Company. Principal partners were Francis F. Palms, president; William B. Moran, vice president; Frederick T. Moran, secretary and treasurer; W.P. Ratigan, and Waldo Johnson. The prominent mining engineer William M. Courtis served as consultant. The Peninsular company set a force of men to work under Captain Thomas Trevithic, sinking the shaft to 69 feet, then drifting and crosscutting to test the extent of the vein. The quartz was found to be 31 feet wide at the bottom of the shaft. A drift at the 50-foot level followed the vein 113 feet, while trenching showed it to continue at least 400 feet on the surface, as well as uncovering four other intersecting veins. A second shaft was sunk south of the first, eventually reaching a depth of 80 feet.

Although samples from the shafts and drifts assayed as high as $120 to the ton and specimens were found containing nuggets the size of peas, a mill test was needed to prove what the mine could actually produce. In March 1891, 40 tons of lean ore, with no gold visible, were run

> AN INVESTOR'S MINE.
>
> DESCRIPTION
> OF THE PROPERTY OF
>
> Peninsular Gold Mining Co.
>
> SITUATED ON
> THE ISHPEMING RANGE,
> IN MICHIGAN.
>
> ITS RESOURCES
> AND FUTURE PROMISE.

through the stamp mills at the Ropes mine, yielding an average of $3 per ton. The mine continued to be worked through the summer, proving up the vein, but despite the promising mill test results, the good assays, and the abundance of ore available, operations ceased in August 1891. The company had failed to attract enough investors to keep the mine running, and the lease was surrendered to Gingrass the following year.

In 1898, Charles T. Fairbairn and Albert K. Sedgwick, who also held an option on the idled Ropes mine, took an option on the Peninsular property. The shafts were dewatered and samples were taken, but not enough gold was found to attract investors to reopen the mine. Another option was taken by parties from the Copper Country in 1901, but apparently no work was done. In 1905, the property was among seven forties taken over by the Ishpeming Gold Mines Company. This company reopened the Michigan as a silica mine, but did nothing with the Peninsular.

J. Maurice Finn

The property was acquired by Callahan Mining Corporation in 1985. After a program of diamond drilling and surface geological study, Callahan announced in October 1987 that it would reopen the mine. Mining would be carried on by a method similar to that used at the Ropes, with a production level at a depth at 400 feet and two or three mining levels at intervals above that. The ore would be hauled to the surface in underground trucks traveling up an inclined tunnel and would be trucked to Callahan's Humboldt plant for processing. After more than 100 years, it seemed that the Peninsular Mine would finally live up to its promise. Further drilling, however, did not prove as promising as the earlier work, and the development project was suspended.

After Callahan left the Upper Peninsula, mineral exploration company Aquila Resources Inc. obtained an option on the Peninsula property and began a diamond drilling program. A total of 43 holes were drilled in 2010-11, which outlined a gold-bearing zone up to 1,000 feet long and 500 feet deep. One drill hole cut through a vein that carried 9.2 ounces of gold per ton – about $2,070 per ton at 2010 prices. The Peninsular property continues to show the potential to become a valuable gold producer.

In the fall of 1888, attorney J. Maurice Finn of Grayling, Michigan, began exploring for gold near the workings of the Peninsular Gold Company. Tracing the Peninsular vein to the west, he found two promising veins of gold-bearing quartz on the S ½ of the SE ¼ of Section 26, T48N-R28W. Shunning publicity, Finn worked to secure a lease on the two forties from the owners, the heirs of the late Senator Edward N. Breitung of Negaunee. Working as general manager for the newly organized Grayling Gold and Silver Mining Company, he quietly started sinking a shaft on the eastern forty. The Grayling Gold and Silver Company was organized by businessmen from downstate Grayling. Rasmus Hanson was the president, Finn served as secretary, and Nels Michelson was the treasurer. Finn's secrecy was broken in February 1889, when a local newspaper announced the find, reporting the rock to be so rich that the hand-driven drill used by the prospectors became stuck in the soft native gold. The first blast broke loose more than 7 ounces of gold, and some specimens of quartz were found to contain

Nels Michelson

LEE DEGOOD

up to $38,000 per ton in gold. The company equipped the property with a steam hoist and built a boarding house, nicknamed the "Hotel Breitung," to house the 12 miners. Finn attempted to engage the Peninsular and Michigan companies with the Grayling in the joint purchase of a five-stamp mill. Failing this, the company proposed to buy a 20-ton capacity Huntington centrifugal mill for its own use, but backed out when the manufacturer would not provide a suitable warranty.

When the shaft reached 60 feet, drifts were started, following the vein to the east and west, and a crosscut to the south sought the other vein that showed on the surface. Specimens rich in native gold were still being found, so the shaft was deepened to 100 feet. In October 1889, the Grayling company contracted with M.E. Harrington and Son of Ishpeming for diamond-drill exploration, the first use of the diamond drill on the Ishpeming Gold Range. The first drill hole struck the vein at a depth of 400 feet, finding pay rock averaging $28.84 per ton. A second drill hole was started 80 rods west, near the center of the property, in 1890. A formation of soft rock that caved into the hole slowed the work and eventually stopped the drilling when the diamond bit and drill rod became stuck 250 feet down. Encouraged by the results of the drilling, the company started a shaft at the second drill hole, but ran into trouble sinking through the same soft formation that had stopped the diamond drill. The company quit work shortly thereafter, having spent more than $15,000 without any return. The next time the Grayling property was mentioned in the press was in 1894, when it was advertised as available for exploration at reasonable terms.

In addition to his explorations for the Grayling Gold and Silver Company, Finn had an interest in several other gold prospects. In partnership with Julian M. Case of Marquette, he bought an option on the S ½ of the SW ¼ of Section 26, T48N-R28W, from its owner, Theodore M. Davis. There he found a vein of gold-bearing quartz in the granite country rock. Assays showed $10 per ton, but the vein was too narrow to be worked. Finn also joined with Ishpeming blacksmith Thomas Gaynor to form the Ishpeming Gold and Silver Exploring Company in 1889. They bought an option on the S ½ of the NE ¼ of Section 23, T48N-R28W, where landowner Joseph Pepin had found a 12-inch vein of quartz, but it is not known how much work was done. J. Maurice Finn left the gold mining business with the closing of the Grayling prospect, but re-entered the public eye when he made an unsuccessful bid for the 12th District congressman's seat in 1892 amid allegations that he had defrauded the state and Nels Michelson, his former partner in the Grayling company.

Finn left Michigan in 1893, leaving a trail of bad debts behind him. He relocated in Cripple Creek, Colorado, where the Colorado gold rush was in full swing. Once again Finn became involved in gold mining. He founded and served as president of the Mountain Beauty Gold Mining Company, a position he also held with the Lady Campbell and Royal Oak Gold Mining companies. He served as secretary of the Little Alice Gold Mining Company and the Amazon Mining Company. Finn also resumed the practice of law, adding a collection agency and real estate business. He soon built a fortune buying and selling gold properties and handling miners' legal disputes. When vice presidential

nominee Teddy Roosevelt visited Cripple Creek during the campaign of 1900, Finn invited him to be his guest the next time he visited the West. Roosevelt accepted, so Finn proceeded to build a palatial 26-room mansion in the best section of town. This grand home, named "The Towers" by Finn but known to the locals as "Finn's Folly," was five stories high and boasted an observatory and an indoor trout pond surrounding its central stairway. When Vice President Roosevelt visited Finn in the spring of 1901, he proclaimed J. Maurice Finn's "The Towers" to be the most beautiful home in Colorado.

The Towers

**The Mountain Beauty Gold Mining Company.**
Incorporated.

Directors
J. Maurice Finn......President and General Manager
A. E. Carlton......................Vice-President
M. W. Levy............................Secretary
H. C. Cassidy...........................Treasurer
Geo. Rex Buckman.
Main Office—365 Bennett avenue, Cripple Creek, Colorado.

The NW ¼ of the SE ¼ of Section 35, T48N-R28W, south of the Michigan mine, was explored by George Cummings in 1888 with unknown results. Cummings was married to Hannah Ropes, the sister of Julius Ropes, and was a partner in the Ropes Gold and Silver Company. He was a highly respected mineral explorer and had discovered the Holyoke silver mine in 1864, but he was a very secretive man. One of his last acts before he died was to burn his life's collection of geological notes, perhaps destroying information on mineral deposits undiscovered to this day.

Late in 1888, Ishpeming merchants Richard and John Mockler found a 3½-foot vein of gold-bearing quartz a mile southwest of the Michigan mine on the SW ¼ of the SW ¼ of Section 35, T48N-R28W. The brothers sank an exploratory shaft to a depth of 25 feet and found free gold assaying up to $17.50 per ton, but not in enough quantity to make the prospect pay. Explorations were limited to the fall and winter of 1888.

Three quarters of a mile west of the Mockler shaft are two shafts and a number of trenches, the explorations of two other early mining companies. Organized in July, 1888, the Lake Superior Gold Mining Company leased the E ½ of the SW ¼ of Section 34, T48N-R28W, where a vein of gold-bearing quartz had been found, from the East Saginaw Iron Company. The company's founders, Richard Blake, Irving D. Hanscom and C.F. Conrad of Marquette, offered stock for sale at 10 cents per share to finance exploration. The company explored by trenching and test pitting and sank a shaft at least 38 feet before giving up the prospect. A year later, a company of Detroit capitalists formed the Detroit Gold Mining Company. Dr. H.O. Walker was president and L.H. Collins served as secretary. The company was reported to hold a very promising forty on Section 34, but it is not known if they took over the Lake Superior company's workings or opened a new prospect of their own.

Although the Michigan Gold Range ran generally east and west through Sections 35 and 36, promising quartz veins were also found on several sections to the north. The Peninsular, Grayling and Finn Case prospects were located immediately north on Sections 25 and 26. Section 23, T48N-R28W, owned by Leon Pepin, was also the site of several explorations in 1887 and 1888, including the previously mentioned Ishpeming Gold and Silver Exploring Company, which was located in the NE ¼ of the section.

In 1888, Dr. A.E. Gourdeau, the county physician, took an option on the SE ¼ of the SW ¼ of Section 23 and set a few men to work exploring. On the adjacent forty to the west a Hungarian immigrant named Smith was working

> **RICHARD BLAKE,**
> DEALER IN
> **Real Estate and Pine.**
> **TIMBER AND MINERAL LANDS.**
> City Business and Residence Property for Sale. Sole Agent for the Longyear Addition to Marquette.
> **HARLOW BLOCK.**

at sinking a shaft into an outcrop of quartz. The shaft reached a depth of 22 feet before Smith sold out. A group of Ishpeming men, led by Frederick Braastad, the owner of the Winthrop iron mine and Moses B. Toutloff, proprietor of the Hotel Toutloff, purchased the Gourdeau and Smith options and began exploring in the spring of 1889. The local men attracted the attention of investors from Grand Rapids and formed the Grand Rapids and Ishpeming Gold and Silver Company. Braastad served as president, Toutloff was treasurer, T. Hughes was the secretary and C.L. Love served as business agent. Smith's shaft was pumped out, and testing of the vein was resumed. Aside from one newspaper article reporting that "fine quartz" was being found, the prospect was never mentioned in the press again.

In the fall of 1888, Ephraim E. Coon and E.P. Bennett, who were involved in the Gitchie Gumie Gold Mining Company, discovered another gold-bearing vein 3½ miles west of the Michigan Gold Mine. Located on the SW ¼ of the NW ¼ of Section 32, T48N-R28W, near Boston Lake, the prospect was explored under option from the East Saginaw Iron Company. A crew of four men tested the vein and reported finding paying quantities of gold-bearing quartz.

Lacking the finances to develop the lode, Coon sold his option to John Paulson of Minneapolis. Along with John Q. Adams and J.F. Foley of Negaunee and Joseph A. Ames, J.W. Edwards and C.A. Avery of Milwaukee, Paulson founded the Ishpeming Gold Syndicate. The syndicate bought Paulson's option. No evidence of Bennett's and Coon's explorations has been found, so it would appear that the prospect did not prove valuable.

## Chapter 7

# Ropes Range Prospects

Located on the 40 acres adjacent to the Ropes mine on the west, the Phillips Gold Mine was typical of most of the prospects on the Ishpeming Gold Range in that it never produced any appreciable amount of gold. The Phillips was unusual, however, because it may have been an outright fraud.

The land surrounding the Ropes Gold Mine was owned by the Deer Lake Iron Company, mainly as a source of hardwood to make into charcoal for the Deer Lake blast furnace. After mining at the Ropes in 1883 began to prove the vein's value, the Deer Lake company, under its manager William H. Rood, began exploring for gold on their lands to the east and west of the Ropes along the strike of the vein. Early in 1884, a vein of gold-bearing quartz 5 feet wide was found about a quarter-mile west of the Ropes shaft on the SE ¼ of the NE ¼ of Section 30, T48N-R27W, and an 8-by-10-foot test shaft was sunk upon it. The miners were still finding rich quartz when water seeping into the shaft stopped the sinking at 40 feet.

Later in 1884, the Phillips Gold Mining Company was formed to purchase and mine the vein on the Deer Lake company's land. Incorporated in Springfield, Illinois, under the laws of that state, the company was founded by portrait painter John Phillips, iron ore capitalist H.R. Dunkee, real estate tycoon Andrew J. Cooper, Roswell B. Bacon and W.H.A. Brown, all of Chicago, and Joseph Sellwood of Ishpeming. The company purchased the property for $40,000 and offered 100,000 shares of stock at $10 each, a price that local newspapers considered unreasonably high for an unproven prospect.

Under Superintendent John Davies, an experienced gold miner from Colorado and one of the company's strongest promoters, the miners dewatered the existing shaft excavated by the Deer Lake company, using the boiler and steam engine from the original Ropes stamp mill. The Phillips company also decided to drive an adit into the side of the hill some 50 feet east of the existing shaft. When the adit reached the nearly vertical quartz vein 35 feet into the hill, a second shaft was sunk, following the vein and intersecting the adit. Assays of samples from the surface of the vein showed $5 to $12 in gold, and at 22 feet in the new shaft the quartz was said to average $36 per ton. The company announced plans to sink the shaft to 150 feet and install a stamp mill to extract the gold, but by 60 feet the vein had either pinched out or angled away from the shaft. The company continued to sink the shaft, searching for a continuation of the vein, and eventually reached a depth of 90 feet in late 1885.

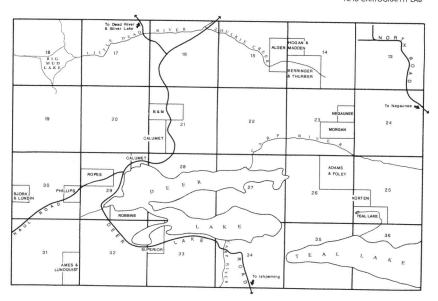

The Ropes Gold Range

NMU CARTOGRAPHY LAB

**Deer Lake furnace**

ALL IMAGES AUTHOR'S COLLECTION UNLESS INDICATED

DEER LAKE FURNACE AND FALLS

While the prospect was producing rich gold samples, Superintendent Davies and Charles Ralston, the local agent for the Phillips company, promoted the stock heavily, both in the local area and in the Chicago and Milwaukee markets. Ambitious plans to build a 25-stamp mill were announced in 1885. When the shaft lost the vein, however, the company withdrew its stock from sale in the local area, and Davies and Ralston were little to be seen. In the downstate markets, the stock continued to be sold at about the same price as Ropes Gold and Silver Company stock.

By March 1886, all work had been abandoned and the company's representatives had disappeared from the area. Taxes on the property were left unpaid after 1888, and the land and mineral rights eventually reverted to the state. Callahan Mining Corporation bought the property in 1983, but didn't see any promise in reopening the prospect – in fact they filled the shafts and built a haul road between them, and buried the adit beneath the waste rock dump of the Ropes mine. Although the mine may have been a legitimate prospect in the beginning, it seems the objective of the Phillips Gold Mining Company was to mine gold from investors' pockets, rather than from veins of quartz.

By the spring of 1889, the Superior Gold and Silver Company had nearly given up on its property adjacent to the Michigan mine. About that time, Julius Ropes found a new gold vein on Section 33, T48N-R27W, just south of the Carp River between the Ropes mine and the Deer Lake Furnace. Ropes and his associates had enough faith in their new prospect to announce plans to install a stamp mill as soon as possible. This inspired the Superior Gold and Silver Company to buy the prospect, comprising the W ½ of the NW ¼ of the section, financed by a 10-cent-per-share assessment. The prospect on the Michigan vein was abandoned, the blacksmith shop and mining equipment were moved to the new prospect and by mid-June a 15-foot shaft had been sunk on the vein. Again, despite vigorous exploration and assays showing up to $179 per ton, economic quantities of gold were not to be found.

Early in 1883, Captain George Berringer, an experienced iron-mining man, found a 6-foot-wide vein two miles northeast of the Ropes on the SW ¼ of Section 14, T48N-R27W, near Zhulkie Creek. Here he sank one test shaft on top of a hill and a second one farther east in a hollow. Encouraged by assays showing up to $30 per ton in gold and silver, and financed by prominent lumberman Henry C. Thurber of Marquette, he continued sinking the shaft to 50 feet, sending the rock to smelters in Chicago and Newark for testing. A

NEGAUNEE HISTORICAL SOCIETY

John Q. Adams

lack of capital forced the mine to be shut down during the winter. The mine reopened in July 1884 with financial backing from a pool of Chicago investors, who confidently made plans to erect a stamp mill. Further work, including sinking the shaft to nearly 100 feet, did not bear out their confidence, and by December, the prospect was closed for good.

West of the Berringer & Thurber prospect, and apparently on the same vein, a group of Negaunee men started a small shaft late in the summer of 1883. Organizing the Negaunee Gold and Silver Company, they leased the SE ¼ of the NE ¼ and the NE ¼ of the SE ¼ of Section 15, T48N-R27W. The miners sank the shaft to a depth of 41 feet and found ore worth up to $50 per ton before ceasing work that fall. The prospect sat idle until the next fall, when the company was reorganized as the Alger Gold Mining Company, named in honor of Michigan's Civil War hero (and later governor) Russell A. Alger. Rock assaying up to $34.90 per ton was found, with nearly half the value in silver. The shaft was worked for about a month, reaching a depth of 60 feet by the end of December. The vein proved inconsistent, however, and work was stopped again until spring. Work was resumed in April, and by the end of August 1885 the shaft was 90 feet deep. Thirty-dollar ore was still being found, but not in sufficient quantity to pay, so the prospect was abandoned. Among the principals in the Negaunee and Alger companies were Messrs. Morse, Dougherty, Davis, J.F. Stevens, Thomas M. Wells, attorney John Q. Adams, *Negaunee Iron Herald* Publisher Clinton G. Griffey, Joseph M. Gannon, iron mining Superintendent James F. Foley, Edward Blake, and Mathias Schweisenthal.

In the fall of 1883, a third prospect was opened on the

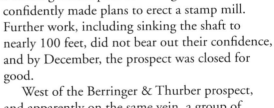

PHILLIP J. HOGAN,
—MANUFACTURER OF—
CARBONATED ✦ DRINKS.
**SAVE YOUR BOTTLES!**
I will purchase Apollinaris or Ginger Ale Bottles at a uniform rate of FOUR CENTS per bottle. Ship by freight at my expense to
**P. J. HOGAN, NEGAUNEE, MICH.**

Berringer vein. Negaunee businessmen Phillip Madden, a saloonkeeper, and Phillip Hogan, who sold "temperance drinks," along with Peter Fitzpatrick and John T. Hayes, started a small shaft on the SW ¼ of the NW ¼ of Section 14, T48N-R27W. In October, they released assay reports showing up to $4.13 in gold and $21.34 in silver per ton. The shaft was sunk about 20 feet before being shut down for the winter. Work was resumed the next August, and the shaft was sunk another 10 feet, with unknown results. No further assays were made public, but Hogan, Madden and Fitzpatrick organized a stock company along with Messrs. Mooney, Molloy and Welsh of Ishpeming. The company announced plans to sink a second shaft 200 feet south of the first and carry it down at least 100 feet, but this was the last mention of the prospect in the local press.

General Russell A. Alger

    The Teal Lake Gold and Silver Exploring Company prospected on Section 25, T48N-R27W and adjacent sections north of Teal Lake in 1885, working under an option from the Michigan Land and Iron Company. The company did not publicize the results of their explorations, but ceased operations a short time later. Officers in the company were A.A. Anderson, president; W.O. Tislov, treasurer; and C.W. McMahon, secretary. A shallow shaft on the north bank of Sucker Creek is all that remains of the Teal Lake Gold and Silver Exploring Company's workings.

    A small shaft can be found on the side of a hill in the NE 1/4 of the NE 1/4 of Section 35, 48 27, half a mile west of the Sucker Creek shaft. The shaft exposes a quartz carbonate vein with pyrite and chalcopyrite. This is probably the prospect of Charles Smith, who was reported in 1888 to have found gold-bearing quartz and galena in serpentine in this section.

    In the spring of 1889, the Korten Gold and Silver Company was formed in Negaunee to explore lands north of Teal Lake, where George Korten had discovered quartz carrying free gold. Peter E. Gingrass and John

John Jochim

W. Jochim of Ishpeming were elected president and treasurer, respectively, and Christopher Roessler and City Attorney James L. McClear of Negaunee were vice president and secretary. The company traced the vein across the NW ¼ of the SW ¼ of Section 25, T48N-R27W, and reported finding specimens carrying gold and silver worth up to $35 per ton. During the summer of 1890, after a change of management, the company stoped into the bluff and sank a shaft at the richest spot on the vein, but failed to find pay rock.

Negaunee experienced another period of gold fever in 1891, when Abram Boulsom discovered a mineralized vein three miles north of town. Boulsom was a merchant tailor who had emigrated from Finland as a young man and plied his trade in Hancock before settling in Negaunee, where he served as alderman from the city's second ward. He had built up a successful tailoring business in the Boulsom Block on Iron Street before selling his shop and going into the gold mining business in 1891. Early that year he and several associates formed the Negaunee Gold and Silver Mining Company to develop a gold prospect he had found on the SE ¼ of the NE ¼ of Section 23, T48N-R27W. Boulsom was the first president, and Theophile Roy was secretary and treasurer. An assessment of 1 cent per share was called to finance work at the prospect. Within a few months, the company's

JAMES L. McCLEAR,
Attorney at Law,
NEGAUNEE, MICH.

ABRAM BOULSOM,
MERCHANT TAILOR.
Imported and Domestic Cloths
Of the Latest Styles.
COMPLETE FITS GUARANTEED.
NEGAUNEE,   -   MICHIGAN.
Orders from adjoining towns solicited and promptly attended to.

directors noticed irregularities in the books and fired Roy, replacing him with new treasurer George Maas and new secretary Alexander Meads.

The vein of gold-bearing quartz at the Boulsom prospect was only 6 inches wide at the surface, but as the shaft followed the steeply dipping formation to 36 feet, it widened to 9 inches. An assay of 5 pounds of rock from the bottom of the shaft showed $182 in gold and as much as 80 ounces of silver per ton. Despite this promising assay, the narrow vein apparently never widened enough to furnish paying amounts of gold, and the work was abandoned that fall with the shaft only 40 feet deep.

LEE DEGOOD

Mining at the Negaunee Gold and Silver Mining Company's prospect was done by contract miners under the direction of James Morgan, who also explored adjacent lands for other gold and silver veins. He found a promising vein a quarter mile southwest of the Negaunee company's shaft and started sinking his own shaft. How deep he dug and what results he had are unknown, but a small, timber-cribbed shaft and a pile of broken rock on the N ½ of the SE ¼ of the section are testimony to his efforts. Morgan also leased the Negaunee company's property in 1894 and reopened the shaft. A crew of three men worked to dewater the shaft and continue sinking, but it soon became apparent that there was little more gold to be found.

Two other shafts were sunk on small gold veins by unidentified parties within a mile of Boulsom's and Morgan's shafts, but results are unknown. Abram Boulsom also explored for iron ore in the Negaunee area, but by 1897 he was back in the tailoring business in

Negaunee in partnership with Charles Peterson, to whom he had sold his business six years earlier.

As the gold boom died down in the 1890s, few prospectors continued to search the countryside for gold. Only the Ropes mine continued to produce the yellow metal, and even it was running on a shoestring.

The promising country between the Ropes and Michigan mines had been explored, and although a number of pits and shafts had been opened, no bonanzas were found. Only a few dedicated prospectors continued to search for riches.

Ishpeming clothing merchant and mineral explorer Edward Robbins was one such prospector. He had explored for minerals in the iron and gold ranges for years. In 1887, he found a small gold-bearing vein, assaying $7.56 per ton, on the S ½ of the SE ¼ of Section 29, T48N-R27W, near what is now known as Gold Mine Creek. Along with tailor George H. Arthur, he explored the E ½ of the SE ¼ of Section 31, T48N-R27W, about a mile southwest, reportedly with promising results. During the fall and winter of 1888, Robbins and Arthur traced the vein, which was 2 to 4 feet wide, some 500 feet on the surface, and reported favorable assays. The next spring, they reported finding a 3-foot vein of quartz carrying pyrite and chalcopyrite

**C. H. KIRKWOOD,**
DEALER IN
**Drugs and Medicines**
CHOICE PERFUMERIES, TOILET and FANCY GOODS.
Preparing Prescriptions a specialty.
MAIN STREET.
ISHPEMING, - MICH.

**ROBERT MAXWELL,**
DEALER IN PLAIN AND FANCY
FURNITURE, WINDOW CURTAINS, WALL POCKETS, PICTURE FRAMES, Etc.

Also GENERAL UNDERTAKER.
Prompt and careful attention given to all orders.
Store and Warerooms, Main Street. - ISHPEMING, MICH.

three miles to the east on Section 34, T48N-R27W. None of these prospects held enough gold to pay, however.

In 1894, Robbins turned his attention to Section 21, T48N-R27W, midway between the Ropes mine and the Berringer and Thurber vein. A 3-foot vein of quartz, bearing silver and gold, had been found in this area by R.D. Vaughn, Robert Maxwell and Charles H. Kirkwood in 1885, but little other exploration had been done. Within a few hundred yards of the charcoal-making location of Ten Kilns and the county highway leading north to the old Coon and Fire Centre prospects, Robbins found a vein of quartz carrying pyrite, chalcopyrite, hematite, and up to $610 per ton in gold. He leased the land from its owner, the Michigan Land and Iron Company, subject to a 10 percent royalty on any precious minerals produced. Although he was still finding rich specimens, Robbins continued to search through 1894 and 1895 for larger veins and found an extension of the vein 2,000 feet to the west. He died in 1896 without opening a shaft to prove his discovery.

Two years later, James H. Billings and William Murdoch secured a lease on the S ½ of the NW ¼ of Section 21, T48N-R27W, where Robbins had made his richest finds. Murdoch, the proprietor of the Murdoch House Hotel, was a retired clown known as "Uncle Billy" who had worked in P.T. Barnum's circus. After retiring to Ishpeming, he continued to entertain

Frederick Begole

the local children on special occasions. Along with Judge Henry H. Mildon and Anson B. Miner, Billings and Murdoch formed the B & M Gold Company. Lumberman Frederick H. Begole was elected president, with Billings as vice president, and Randall P. Bronson as secretary treasurer. The company, financed by its incorporators, started sinking a shaft, following the vein to 20 feet by March 1898. By the time the shaft reached 50 feet, however, the vein had narrowed. The company drifted some 37 feet, optimistically expecting to strike "some pretty good looking quartz" when they hit the "main vein." The main vein was never found, and the company ceased operations late in 1898.

In 1935, Elmer Strathman leased the B & M property from the Ford Motor Company, which had bought out the land holdings of the Michigan Land and Iron Company. Inspired by the reopening of the Ropes and Michigan mines, as well as the increase in the price of gold to $35 an ounce, he started a one-man mining operation at the old shaft. He built a grinding mill from a steel drum and water pipe, and set up a hoist powered by an old truck chassis on the waste rock pile in front of the hillside shaft. Strathman apparently had no better luck finding the elusive main vein than the B & M company had.

A pair of small shafts located half a mile south of the B & M shaft are the only remaining traces of the Calumet Gold and Silver Mining Company. In the fall of 1883, the Calumet company took an option on the NW ¼ of the NW ¼ of Section 29, T48N-R27W, just north of the Ropes. The prospect was disappointing, however, so the company turned its attention to the W ½ of the SW ¼ of Section 21, T48N-R27W, which it also held under option. The miners worked into the winter sinking two shafts 100 feet apart, but failed to find pay rock. The Calumet Gold and Silver Mining Company was founded by John Quincy Adams of Negaunee, president, Joseph Sellwood of Ishpeming, vice president, Joseph C. Foley of Chicago, secretary/treasurer, and James F. Foley of Negaunee and D. Kloeckner of Hancock, directors.

Elmer Strathman at the B & M prospect
SUPERIOR VIEW

During the summer of 1894, James Ames and Peter J. Lundquist of Ishpeming, who had discovered the Ames iron mine east of Ishpeming the previous year, began prospecting for gold. After unsuccessfully exploring a forty adjacent to the Peninsular mine, they moved their operation to Section 31, T48N-R27W, near Edward Robbins' earlier find. When they discovered a 6-inch vein of gold-bearing quartz, they leased the south half of the section from the Cleveland Cliffs Iron Company and sank a shaft. Ames and Lundquist mined several tons of ore and reported assays of up to $50 per ton, but did not find enough rich ore to make the prospect pay.

On Section 18, T48N-R27W, some two miles north of the Ropes mine, Anson B. Miner, one of the discoverers of the Michigan mine, reportedly found gold. Located during 1885, the well-defined vein produced some good specimens. Details of the location and value of the vein are lacking.

The Anglo American Land and Mineral Company was formed in 1885 by W.A. Allen, R.H. Taylor and Byron Jones of Negaunee. The company was organized to develop gold and silver properties in the Marquette

County area, but the location of the company's lands is unknown.

Richard Crow, a highly experienced mining man from Boulder City, Colorado, (today's Boulder) was the mill superintendent at the Ropes mine. Recognizing the possibility of further gold veins being found nearby, he joined with Anson Miner and Joseph Sellwood, the general manager of the Ropes mine, in exploring to the east on Section 27, T48N-R27W. Crow sank a 10-foot shaft into a small vein late in 1884 and produced a small amount of rock worth from $5 to $18 per ton.

About the same time, James F. Foley and John Quincy Adams of Negaunee, partners in the Hematite Mining Company, found gold-bearing rock two miles east of the Ropes mine. Located on the NE ¼ of Section 26, T48N-R27W, the prospect yielded specimens carrying up to $34 in gold and silver, but was abandoned by 1885.

Twelve years later, in 1896, Charles T. Fairbairn, William Brooks and J.O. Flack resumed explorations on the same section. The men found a vein up to 10 feet wide and several hundred feet long, but assays of rock from their 16-foot shaft proved disappointing. Fairbairn continued to explore the area, however, and in 1900 found a small vein of quartz carrying as much as $48 in gold to the ton, but no mining was done.

In 1936, a group of investors that included Houghton attorneys Allen F. Rees and Dean Robinson, incorporated the Norgan Gold Mining Company. The company conducted extensive explorations across the gold ranges on lands optioned from the Ford Motor Company, which had bought up the holdings of the Michigan Land and Iron Company. Late in 1936, Norgan's prospectors found a vein of quartz carrying chalcopyrite on the SW ¼ of the NW ¼ of Section 30, T48N-R26W, north of Teal Lake near the Carp River. Samples from the 3-foot-wide vein assayed up to six-tenths of an ounce per ton in gold ($21 at the new price of $35 per ounce). The following year, the company sank two shallow shafts on the vein, but apparently failed to find paying quantities of gold.

# CHAPTER 8

# The Dead River Gold Range

Once the silver boom of the 1860s had died down with the closing of the Holyoke mine in 1868, this northern mineral range was largely ignored for years. There were sporadic attempts to organize companies to reopen some of the silver lead mines, but little was done beyond exploration. The potential for gold deposits in northern Marquette County was seemingly forgotten, despite the numerous minor gold finds that had been revealed by the silver mining companies, including the first discovery of gold at the Marquette Silver Mining Company's prospect near the Rocking Chair Lakes.

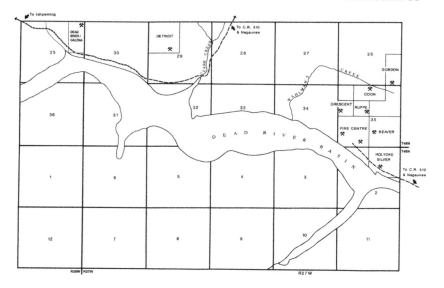

The Dead River Gold Range
NMU CARTOGRAPHY LAB

137

With the discovery of commercial amounts of gold at the Ropes Gold Mine in the 1880s, people began prospecting for gold all across Marquette County. Some eight miles southeast of the Marquette Silver Mining Company's prospect, Charles Ruppe, along with Joseph Loranger and M. Carey, found a vein of quartz carrying gold and copper ore in 1884. Ruppe's prospect was in the NW ¼ of Section 35, T49N-R27W (the same quarter section where the Fire Centre Mining Company later opened the Crescent prospect). Over the next three years, Ruppe sank two shafts and reported finding ore worth up to $30 per ton. Located near the head of a long valley, the shafts were 10 and 32 feet deep.

Nearly half a mile to the north, at the foot of the valley, Ephraim Coon of Ishpeming found an extension of the Ruppe vein in 1884. Coon optioned the NE ¼ of the NW ¼ and the NW ¼ of the NE ¼ of Section 35, T49N-R27W and started work. Shaft sinking and testing revealed ore containing 25 to 50 percent copper and gold worth up to $39 per ton. In July 1885, the Coon Gold Mining Company was incorporated. Ishpeming's mayor, Christian Melby, was elected president, with Montgomery Thompson as vice president and Cornelius Kennedy as treasurer. The company cut a wagon road to the mine from the Ten Kilns location north of Ishpeming and spent the summer prospecting for additional veins. Further shaft sinking was hampered by water entering the shaft, which was located at the bottom of a valley, only a few feet away from a seasonal stream, which would grow from a trickle to a raging torrent after every storm. With financial help from Copper Country investors, a boiler and steam pump were purchased and hauled to the mine in March 1886, and mining was resumed. With the spring snowmelt, however, the shaft, which had reached a depth of 40 feet, was

ALL IMAGES AUTHOR'S COLLECTION, UNLESS INDICATED

**C. MELBY & CO.**

DEALERS IN

**DRY GOODS,**

Boots and Shoes, Groceries,

GLASSWARE, FLOUR, FEED,

and General Merchandise.

100 SOUTH MAIN.

flooded again. The steam pump proved inadequate to keep the water down, and the prospect was abandoned. Today, no sign remains of the shaft except the waste rock dump and the iron pipe from the steam pump. The shaft itself has been completely filled by sediment washed down the valley.

North of the Ruppe and Coon prospects on the Dead River range was a prospect started in September 1885 by a group of men from the Copper Country. Headed by lumberman John R. Gordon, they explored across a wide valley from the Coon on Section 26, T49N-R27W. On the E ½ of the SE ¼ of the section, they found a gold and silver-bearing vein, which they explored from an adit driven into the side of the hill. By the end of October, the adit was 35 feet into the 8-foot-wide vein, and an assay showed $8.56 per ton in gold and silver. At a meeting in January 1886, the investors decided that the prospect was rich enough to warrant further exploration and ordered that a shaft be sunk on the vein. The gold-bearing quartz was reported to have increased in width and richness as the depth increased, and in March, Gordon announced plans to expand operations. Arrangements were made to build a new camp, hire more men, and equip the prospect with a boiler, air compressor and power drills, and a contract was let to sink the shaft another 50 feet. Gordon met later that month with investors from Chicago, hoping to interest them in the mine, but apparently had no luck. Work was abandoned some time later, the gold ore (and perhaps the working capital) having proven too scarce.

Inside looking out from the adit at the Gordon prospect (at the entrance is the author's prospecting dog, Nikki)
DAN FOUNTAIN

In 1888, John Sanson, the discoverer of the Peninsular Mine, prospected north of the Dead River between Silver Lake and the old Holyoke silver mine. Near the northeast corner of Section 25, T49N-R28W, he found a 4-foot quartz vein carrying galena with silver and gold. Sanson obtained an option on the E ½ of the NE ¼ of the section from its owner, the Michigan Land and Iron Company. He soon sold his option to John McDonald of Iron River, who along with investors from Ishpeming, Chicago, and Oconto, Wisconsin, incorporated the Dead River Gold Mining Company. McDonald sold his option to the company for 13 percent of the new company's stock and became one of the directors. Other major stockholders were Henry W. Walker of Chicago, William B. and Sarah F. High of Oconto, and James H. Molloy of Ishpeming. The company set two shifts of men to work, and eventually sank three or four shallow shafts.

> **CORNELIUS KENNEDY.**
> **JUSTICE OF THE PEACE.**
>
> Collections made and remittances promptly returned. Conveyancing and other legal work correctly and promptly attended to. Postoffice box 250.

It seems that the prospecting shafts turned up more silver than gold, since in August McDonald formed another company, the Galena Silver Mining Company, and transferred the option to them. Founders of the new company were Austin W. Wright, Henry S. Heth, and George R. Ghiselin of Chicago, as well as John D. Campbell of Negaunee and McDonald. In the fall of 1889, McDonald was showing off specimens of gold-bearing quartz from the prospect and claiming that he had the richest gold mine on earth. This "richest gold mine" was never again mentioned in the press, however, so it may be assumed that no more pay rock was found.

With the resurgence of gold prices in the 20th century, many abandoned prospects were again explored for gold and silver. Callahan Mining Corporation conducted a diamond drilling and surface exploration program near the workings of the old Dead River Gold and Galena Silver Mining companies and found gold mineralization in areas of brecciation and quartz/albite/

dolomite alteration, as well as within gold-bearing quartz veins. After Callahan closed the Ropes Gold Mine and left the U.P., Aquila Resources acquired the mineral rights to the Silver Creek property and drilled a series of diamond drill holes in 2011. This drilling revealed a gold-bearing zone about 500 feet long and up to 80 feet thick. The prospect remains a potential producer.

By the fall of 1890, John McDonald was promoting another gold prospect a mile and a half to the east. He was able to interest investors from Lower Michigan and Wisconsin and, in October 1890, organized the Detroit Gold and Silver Mining Company. J. VanKirk of Janesville, Wisconsin, was elected president, with Frederick F. Campau of Detroit as vice president and Alex Richardson of Janesville as secretary. McDonald and Edward Campau served as directors.

The new company leased the NW ¼ of Section 29, T49N-R27W from the Michigan Land and Iron Company and started driving an adit into the southern point of a hill, just above the level of the plain. By October 1890, the adit was 15 feet into the hill on an 18-foot vein of gold-bearing quartz. Assays of the vein rock ranged from $2.48 to $428.28 per ton, and the adit was continued 50 feet into the hillside. The vein was exposed at four other points on the property, and a small shaft was started at the richest showing, up a small valley to the northwest of the adit. The shaft reached an undetermined depth before the Detroit Gold and Silver Mining Company's prospect closed down.

Inside Detroit Gold and Silver Company's adit
DAN FOUNTAIN

Years later, the Detroit adit served quite a different purpose. During prohibition, some enterprising Ishpeming men found the isolated mine tunnel to be an excellent location for a moonshine still. The men would haul the ingredients in on their backs several miles from the nearest road and cook up their homebrew inside the adit. Remnants of the still can still be found at the site.

In 1890, Charles Ruppe resumed his explorations on Section 35. Along with tailor Charles Kobi of Ishpeming and George Weatherston of the Ropes mine, he obtained a mining option on the S ½ of the NW ¼ of Section 35, T49N-R27W not far from his old prospect. Ruppe and Kobi soon sold their interest in the option to the Superior Gold and Silver Company, which had two unsuccessful prospects on the Michigan and Ropes Gold ranges, for 1,500 shares of the company's stock. The Superior company sank several test pits and a shallow shaft, but had no better luck here than they had at their previous prospects, and little gold was found. The prospect was apparently shut down the same year.

In 1890, Julius Ropes, who had discovered the Ishpeming Gold Range, undertook a systematic study of the country north of the Dead River, and identified another, separate range of gold-bearing mineral deposits. Stretching from near Lake Superior to Silver Lake, the range was described by Ropes as consisting of knobs and ridges of granite, diorite and greenstone laced with quartz veins carrying free gold along with silver, pyrite, copper ore, galena, zinc ore, and the only reported occurrence of tellurium in Michigan.

Ropes explored lands in the Dead River area, which were held under option by the Dead River Miners' Pool during the summer of 1891. The land was owned by the Michigan Land and Iron Company, which was to collect a 10 percent royalty on any minerals produced. This was later reduced to a more reasonable 5 percent. Ropes's most promising find was on the SE ¼ of Section 35, T49N-R27W, where he found a 3-foot vein of quartz

---

**CHAS KOBI,**
**MERCHANT TAILOR**

First-class Workmanship and a Perfect Fit Guaranteed.
Repairing done.                    116 N. First.

carrying galena and pyrite along with silver and gold assaying up to $502 per ton. This prospect was named the Beaver exploration for the beaver dams on a nearby creek. Two shifts of miners worked to sink the shaft to a depth of 18 feet before a forest fire on July 11, 1891, destroyed their camp and forced the men to jump into the Dead River to avoid the heat and flames. During the next two months, the log camp was rebuilt and the shaft was sunk to 50 feet, at which depth drifts were started to the east and west.

In August, 1891, members of the exploring pool, including Solomon S. Curry of Ishpeming and Charles F. Pfister, Hyatt S. Haselton and George A. Masser of Milwaukee, incorporated the Fire Centre Mining Company, naming it for the igneous origin of the rock of the Dead River range. Curry was elected president, and Haselton served as secretary. The company continued work at the Beaver shaft and started a second shaft just over the hill to the south at a point where quartz holding free gold was found. This became known as the South Beaver lode. Ropes, in his capacity as chief prospector for the Fire Centre company, also located a new vein half a mile to the west. Located just south of Silver Mine Lake, the new find was named the Crescent prospect. This vein, on the side of a hogsback hill, was worked as an open cut rather than by sinking a shaft. An adit was driven into the side of the hill to search for additional veins.

In 1892, some 254 tons of ore from the Fire Centre prospects were hauled by wagon to the Ropes Gold Mine

Crescent adit (with Nikki peeking inside)
DAN FOUNTAIN

**Ruppe shaft**
DAN FOUNTAIN

and treated in the stamp mill. The ore proved to be easily stamped and of good grade, producing $2,063.69 in bullion, an average of $8.12 per ton. This successful mill test encouraged the company to continue to develop the mine and to install a small Crawford centrifugal ball mill capable of milling 10 tons per day. When the mill was started up that fall, however, it was found to be unsuitable for the hard ore, wearing out completely within a month. Representatives from the manufacturer repaired the mill early in 1893, but the mine closed for the rest of winter a short time later. By this time, the South Beaver shaft had reached a depth of more than 120 feet. The North Beaver shaft was at least 50 feet deep and had more than 100 feet of drifts.

Although the Fire Centre company announced its intention to resume operations the next spring, the mines were never worked again. Perhaps the gold-bearing veins had pinched out at depth. More likely, the stockholders were unwilling or unable to sink any more cash into the speculative gold exploration business in the face of the Panic of 1893. The mill machinery was taken to the Ropes mine where the Crawford mill was given another test run in July 1893, again proving a total failure.

The Dead River range was still considered a promising area, and prospectors continued to seek gold and silver in the hills north of the river. The Fire Centre property was optioned in 1903 by unidentified parties, but no mining was done. In 1914, the property was optioned by John Doelle of Houghton, but again nothing was done beyond exploration. During the summer of 1936, the Norgan Gold Mining Company set crews to exploring Section 35, T49N-R27W, starting with surface trenching. That fall, the company rafted a diamond drill across the Dead River basin and drilled 13 holes totaling 2,000 feet. A few rich areas were found, but the veins were not persistent, and the property was abandoned.

# Chapter 9

## Other Marquette County Prospects

The most significant finds of precious metals in Marquette County were clustered in a few areas near Ishpeming, but gold and silver veins have also been found in other scattered areas around the county.

In the course of the U.S. Geological Survey of public land in Michigan in 1847, Dr. John Locke took note of a ridge of metamorphic slate near the Carp River Forge. He reported that the rock formation hosted quartz veins carrying "copper pyrites" and "green carbonate of copper," and recommended that the veins be tested for gold, silver and other rare metals.

Seventeen years later during the silver lead era, explorer George P. Cummings did just that and found enough encouragement that he and several associates incorporated the Sedgwick Mining Company and bought 160 acres in Section 28 and 33, T48N-R26W. Incorporators of the company included Cummings, Henry R. Mather, Sydney E. Church and Silas C. Smith of Marquette, Henry Stewart of Negaunee, and Edwin Parsons of New York City. Mather and Church served as president and secretary/treasurer, the same positions they held in the Holyoke Mining Company.

The company started sinking a shaft on a vein of

---

**GEO. P. CUMMINGS,**

Civil Engineer :-:

——AND——

:-: Real Estate Agent.

Marquette, - Michigan,

quartz which carried abundant chalcopyrite – the "copper pyrites" mentioned by Dr. Locke. Although assays found only trace amounts of gold, the ore was rich enough in copper that the company shipped a 1,000-pound barrel of ore to a smelter for testing. The yield of copper and gold was not enough to pay expenses, so the prospect was abandoned and, in 1883, the property was sold to settle the company's debts. Sedgwick Mining Company director H.H. Stafford was the successful bidder on the land, but didn't do any more work on the vein. In 1888, the Ishpeming Gold Syndicate obtained an option from Stafford to explore the Sedgwick property, but did little work beyond assaying a few samples from the vein before allowing the option to lapse.

## H. H. STAFFORD
DEALER IN
## DRUGS, MEDICINES
**FANCY GOODS, BOOKS, STATIONERY,**
Prescriptions carefully prepared Day and Night.
Cor. Front and Spring Sts.     Marquette, L. S., Mich.

ALL IMAGES AUTHOR'S COLLECTION, UNLESS INDICATED

The rugged hills near the Morgan blast furnace east of Negaunee attracted a number of prospectors over the years. Silas C. Smith, a pioneer prospector, found a vein of silver lead at an undisclosed site in this area in the mid-1860s. The vein was reported to run $70 per ton in silver, but Smith was unable to obtain a lease from the owner of the land, the Iron Cliffs Company, so he abandoned the prospect. Some 20 years later, the aged Smith enlisted the help of Peter White to persuade the iron company to grant them a lease. White succeeded, but Smith died before he was able to relocate the site. White hired James E. Jopling of Ishpeming to head a crew of explorers in prospecting the area. Jopling found a 10-foot vein containing lead ore near Bruce, a stop on the Duluth, South Shore and Atlantic Railroad near Bagdad Pond. Assays of the vein rock revealed only a trace of silver, however, and the prospect was again abandoned.

With the gold excitement stirred up by the production at the Ropes Gold Mine in Ishpeming in the early 1880s, prospectors searched the Morgan area for gold veins. Late in 1884, Negaunee Justice of the Peace John Jones,

Edward Breitung

butcher Charles Muck and merchant Gottlieb Sporley discovered a vein of rich copper ore not far from Morgan. The vein was reported to range from 14 inches to 6 feet in width. Assays showed as much as 43 percent copper, along with $3 to $19.38 in gold and from a trace to $21 in silver per ton of ore. Through the summer of 1885, three men worked at sinking a shaft into the vein, eventually reaching a depth of 18 feet. The partners apparently sold their prospect to state Senator Edward Breitung that fall, when a local newspaper noted that "Hon. Edward Breitung has a prospect near Morgan, which may turn out to be anything from a copper mine to a gold mine." It seems that it became neither, as no further mention of the prospect was made in the press. A shallow shaft sunk on a quartz vein in the SW ¼ of the SW ¼ of Section 25, T48N-R26W, a few hundred yards northwest of the old Morgan Heights Sanatorium may be the legacy of this prospect.

When the Lake Superior and Ishpeming Railroad was built in 1896, it was routed parallel to the DSS&A Railroad through the Bagdad gorge between Morgan and Bruce. The new railroad crossed the Morgan Creek valley on a high trestle next to the abandoned Morgan charcoal iron blast furnace. Between the trestle and the gorge, several small rock cuts were excavated. Either while making these rock cuts or while digging foundations for the trestle legs, the railroad gang found a vein of quartz that held some amount of gold. The value of the vein was not enough to justify relocating the railroad, so it was ignored and the construction crews moved on.

About three-quarters of a mile north of Sporley's prospect is a prospect with a curious history. In 1920, the Beaver Board Company of Buffalo, New York, incorporated a subsidiary, the Beaver Granulith Company, to produce crushed greenstone for use in the manufacture of composite roofing shingles. The company leased the NW ¼ of the NW ¼ of Section 25, T48N-R26W from the Cleveland

Cliffs Iron Company and started an adit into a ridge on the north side of the Bagdad gorge. The rock excavated from the adit was used to build a ramp to the top of the ridge, and an extensive mill was erected on the side of the hill. The company erected several substantial concrete structures, including a powerhouse, machine shop, offices and a modern change house with showers for the workers.

A railroad siding was brought in from the nearby DSS&A Railroad main line. Since the dark-green country rock was the intended product of the mine, it seems odd that the company chose to use expensive underground mining methods rather than simply quarrying the rock from the steep side of the ridge. Perhaps the company hoped that the adit would intersect one of the gold- or silver-bearing quartz veins that were known to exist in the area. Whatever the company's aims, the operation was short-lived. After little more than a year, and with the adit scarcely 50 feet deep, the company gave up its lease, removed the mill equipment and abandoned the mine. Perhaps the rock did not prove to be suitable for roofing, or maybe the adit failed to uncover the gold and silver veins that may have been the true objective of this venture. The Beaver Granulith Company was eventually dissolved in November 1923.

From inside the Beaver Board adit
DAN FOUNTAIN

During his exploration of mineral lands in the 1860s, George P. Cummings discovered numerous mineral deposits, including the Holyoke silver mine and the Baraga County graphite deposits. In many cases he was prospecting professionally for landowners, but he often found promising mineral deposits on government land and was able to buy the properties for himself. In 1864 he found a vein of argentiferous galena near the headwaters of the Chocolay River on the banks of

Gad Smith

what became known as Silver Lead Creek and bought the E ½ of the SE ¼ of Section 30, T46N-R24W. Although the work was not publicized, Cummings apparently tested the vein extensively, as Carl Rominger of the Michigan Geological Survey noted 14 years later that he had seen "many tons of pure lead-ore piled up on the sides of the test pits."

The vein was explored again in 1887, with a group of men from Negaunee working the prospect, described as being five miles from Sands Station on the Chicago Northwestern Railroad. The next time the silver lead vein is known to have been worked was in 1896, when the Michigan Commissioner of Mines and Mineral Statistics reported that Gad Smith, a well-known Marquette politician, and John Miller of Escanaba were sinking a shaft on a vein of galena.

Smith and Miller's shaft was intended to replace a shaft that had been sunk some time earlier by iron mining pioneer Richard L. Selden and partners, but which had been partially collapsed by a poorly placed dynamite blast. The original shaft was reported to be 80 feet deep on a 3-foot vein of galena. When the new shaft reached the old workings, the vein was found to be only 18 inches wide and assays showed "some silver," and the prospect was abandoned.

Two years later, Miller bought out Smith's share of the mine and announced plans to reopen the workings, now described as a 4-foot vein with assays up to 20 ounces of silver per ton. This was the last time the vein was mentioned in the press. The prospect was forgotten until the 1980s, when the county mine inspector determined the partially filled shaft, which was next to a fishermen's foot trail along Silver Lead Creek, to be a safety hazard. The landowner filled it in, bulldozing the silver ore and waste rock back into the shaft.

Ever since the first settlement was established in the 1840s, Marquette was best known as a shipping port for the rich iron mines of Ishpeming and Negaunee.

Beginning in its early days, however, Marquette was the site of a number of gold prospects. In 1854 and 1855, Silas C. Smith discovered mineral deposits that carried from $10 to $140 in gold somewhere within the present city limits. Smith had earlier founded a whetstone business, with a quarry in Negaunee and a factory in Marquette (giving Whetstone Brook its name). He later gained fame for his discovery of the Lake Superior silver lead range near Silver Lake. His gold discoveries in the city were apparently too small for profitable mining and were never developed. Iron ore continued to be the main economic mineral of the area.

In 1876, a deposit of low-grade iron ore was found west of Marquette, not far from the present Marquette Golf and Country Club. Terrence Moore, president of the Ontonagon, Superior and Mammoth Silver Mining companies, which had been mining silver on the Iron River range in Ontonagon County, farmed the land adjacent to this discovery. Moore decided to explore his land for a continuation of the vein of iron. Although the iron ore did not extend onto his land, Moore did find a vein of slate which carried pyrite and chalcopyrite. When he showed Silas Smith a specimen of this rock, Smith brought out a sample of rich gold ore from a mine out west. The minerals were nearly identical, prompting Moore to submit samples for assay to Julius Ropes and to an assayer in Pittsburgh. The assay reports came back showing $6 to $25 in gold and silver per ton.

---

**SAMUEL E. BYRNE,**

## ABSTRACTS OF TITLE

THE ONLY ABSTRACTS OF
Marquette, Schoolcraft and Baraga County Lands.

**MARQUETTE, L. S., MICHIGAN.**

---

Moore quietly proceeded to test the prospect, eventually sinking a shaft 20 feet into the deposit. Finally, in May 1883, when the opening of the Ropes Gold Mine in Ishpeming had stirred up an interest in gold mining, he decided to organize a stock company to develop the mine. Moore served as treasurer of the new Marquette Gold and Silver Company, with Samuel E. Byrne as president, John Buchanan as vice president, Albert Hornstein as secretary, and Thomas Meads as director. The company placed 1,000 shares of stock on the market at 50 cents each to finance development work, and on June 4, 1883, a force of men went to work in the shaft. The directors planned to sink the shaft to at least 70 feet, then start drifting and crosscutting in the vein. They also planned to erect a stamp mill, to be run by water power from a nearby stream. The only news ever published about the prospect following these optimistic plans was in July 1883, when it was reported that the miners were still working, and that the vein was becoming wider with depth. Either gold or capital failed to materialize, however, and the mine faded into obscurity.

What is most likely the site of the Marquette Gold and Silver Company's gold mine is located in a conservation area of the Harlow Farms community. Two depressions with piles of broken quartz next to them are all that remain to show the location of the shaft and a test pit.

Terrence Moore had served Marquette as an alderman and postmaster. He was the founder of several gold- and silver-mining companies, and later became superintendent of the Republic iron mine. His obituary notice in 1901 had the simple directness of the times: "Terrence Moore, formerly postmaster of Marquette and a wealthy and prominent citizen, suicided in Detroit, where he had been living in abject poverty brought on by drink and bad company."

The next chapter in the history of Marquette as a gold-mining town began in the fall of 1884. While grading a street for a new residential area near what is now the corner of High Street and Hewitt Avenue, Captain Smith Moore discovered a deposit of slate and granite laced with 2- to 4-inch stringers

The *Smith Moore*
PETER WHITE PUBLIC LIBRARY

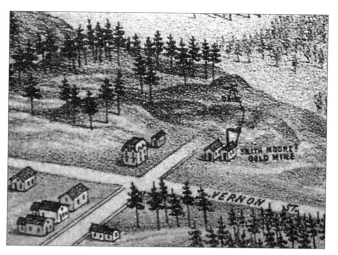

Smith Moore's mine
PETER WHITE PUBLIC LIBRARY

of quartz holding copper ore, iron ore and pyrite. When specimens from the vein were tested, they were found to carry from $4 up to $84 in gold per ton, with traces of silver. Moore, a Great Lakes freighter captain, land developer, and owner of the European House hotel, took a 60-day option on the land from its owner, Peter White, and set a crew of men to work stripping and crosscutting the vein. One of the workmen caused a momentary stir when he uncovered a 30-pound nugget, but when brought to the surface and examined, it turned out to be float copper.

Captain Moore systematically explored the vein, making numerous assays and sending a test lot of the ore to a smelter in Chicago. Most assays came back in the $5 to $10 range, encouraging him to organize the Euclid Gold Mining Company in late December 1884. A block of stock was offered for sale at 50 cents per share and was soon sold out. By the following May, a shaft had been sunk 18 feet at the prospect, and specimens assaying up to $160 had been found.

Richard Crow, the mill superintendent from the Ropes Gold Mine and a highly experienced gold-mining man, inspected the prospect in 1885 and expressed his optimism for the development of a paying mine. Moore shared his optimism, and announced plans to install

Smith Moore owned the European Hotel in Marquette and the steamer that bore his name.
PETER WHITE PUBLIC LIBRARY

either a stamp mill or a Wiswell pulverizer to recover the gold from the rock. Investors had less faith, however, and capital was hard to come by. Captain Moore was forced to seek financial backing outside the local area, and in October he met in Cleveland with investors, who financed the purchase of a steam-driven air compressor and power drill. The new equipment allowed the prospect to be opened more rapidly, but no richer or more abundant ore was found.

In March 1886, a large sample of rock was sent to Chicago for treatment by a new gold-recovery process. Although the results of the test were promising enough to bring a group of Chicago investors to Marquette to negotiate a deal, little more work was done, and the prospect was soon abandoned. Moore never lost faith in the mine, however, and in 1888, he was still attempting to raise the funds to open the prospect. That fall the Euclid company called in an assessment of 5 cents per share on its stock, but little money was paid in, and no more exploration was ever done. Rock from the mine, including much that bore some gold, was hauled away and used in the construction of foundations in Moore's

Addition. As the city grew, the outcrop that hosted the mine was quarried and crushed for road gravel, so newly named Prospect Street and others in the area really were paved with gold!

The shaft and pits were filled in and homes were built where the boiler house once stood, and Smith Moore's gold mine became no more than a page in Marquette's history.

A couple months after Captain Moore's discovery, Marquette grocer John W. Spear took an option and opened a prospect a few hundred feet west of the Euclid Company's. Along with lawyer Dan H. Ball and pharmacist Henry H. Stafford (Julius Ropes's former employer), Spear organized the Marquette Gold Mining Company, with Ball serving as president and Spear as secretary and treasurer. Assays showed $24 per ton in gold and silver, so the company let a contract to sink a 30-foot shaft.

Dan Ball

**BALL & BALL,
ATTORNEYS-AT-LAW**
206-207-208 NESTER BLK.,
MARQUETTE, MICH.

Almost as soon as the company was organized, the founders became embroiled in controversy. Working without the other directors' knowledge, Spear advertised Marquette Gold Mining Company stock in Detroit and Chicago, and sold a large number of shares at 50 cents apiece. As soon as they became aware of Spear's actions, President Ball and Director Stafford resigned, wishing to avoid any association with a possible stock swindle. Stafford and Ball felt that only a limited amount of stock

LEE DEGOOD

should have been sold, just enough to raise working capital to start mining. They hoped that a rich showing as the mine was developed would enhance the value of the stock and strengthen the company's financial position.

As the shaft was sunk during the winter of 1885, the ore became steadily richer. At 25 feet, samples showed up to $27 per ton in gold and silver. When the shaft reached 30 feet, a crosscut was begun, heading south into the footwall. Ore assaying up to $138 was found, but the rich quartz was only found in narrow stringers. Although the force of miners was doubled in the spring and rich specimens were still being found, not enough ore could be mined to treat profitably. By the end of the summer, work at the prospect had ceased.

Dan Merritt

Although he was disappointed by his former partner's business practices, Henry Stafford retained his faith in Marquette's potential for gold mining. In March 1885, he joined with Richard Blake (who was Dan Ball's bookkeeper) and a Mr. Davis in an exploration a few hundred feet northwest of the Moore prospect. The men found an 8-inch quartz vein which yielded up to $31 in silver and $18 in gold, and organized the Eureka Gold Mining Company. In May, they set 10 men to work sinking a shaft on the vein, planning to go to 60 feet. By the end of the month, they had reached a depth of at least 10 feet, and the vein had widened to 2 feet. The rich surface showings must not have continued at depth, however, because the prospect was soon given up.

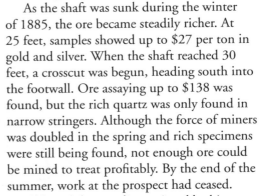

Late in 1885, Marquette County Sheriff Andrew A. Anderson and Ferdinand Bending found a vein of mineralized quartz north of the city. The prospect, located east of the Lake Superior Powder Company's mill along the Dead River, produced specimens yielding $12 per ton in gold and silver. Anderson and Bending began sinking a shaft, following the vein as it widened from 2 feet at the surface to 7 feet at a depth of 25 feet.

With all the gold mines and prospects being opened up in Ishpeming and Marquette, Daniel H. Merritt of the Iron Bay Foundry saw an opportunity to use his metallurgical expertise to join in the gold boom. In partnership with Captain Bailey and Professor Henry, Merritt established an experimental smelter at the foundry. The "gold saver" was tested on gold ores from the Lake Superior prospect in Ishpeming and from Sheriff Anderson's shaft and reportedly recovered up to 93 percent of the gold in the ore, a great improvement over the 75 percent recovery rate for the Cornish stamp mill. Despite these promising results, the smelter never progressed beyond the experimental stage.

Sheriff Anderson's gold mine likewise never achieved success. The shaft reached a depth of only 35 feet before the mine closed down late in 1885. The prospect lay forgotten until the 1970s. When the city of Marquette was developing the baseball fields at the Kaufman Sports Complex, an odd-looking depression was noticed nearby. Excavating down to bedrock revealed a rectangular shaft. To eliminate the hazard, the wood and debris that had filled the shaft over the years were dug out and stable fill dirt was dumped into the shaft and leveled.

One of the shortest-lived prospects in Marquette was begun by Ole Oleson, an employee of the Burtis Sawmill. Early in June 1885, Oleson started a small shaft along Lake Street near the Grace Furnace. Assays

---

ESTABLISHED 1867.

**Rothschild & Bending,**

WHOLESALE DEALERS IN

Imported and Domestic

**WINES, LIQUORS AND CIGARS.**

FRESH IMPORTED AND KEY WEST CIGARS CONSTANTLY RECEIVED.

---

MINES, MINERAL AND TIMBER LANDS

Bought and Sold.

**MINES IN MARQUETTE, MENOMINEE and GOGEBIC IRON DISTRICTS**

For Sale.

Options for Mining Leases secured on Desirable Lands.

EXPLORATIONS FOR IRON ORE CONDUCTED.

Special attention given to

→ Selecting Iron Lands and Exploring for Iron Ore ←

**JULIAN M. CASE,**

Marquette, Mich.

from the prospect at the base of the hill showed some gold and silver, and Oleson optimistically expected to find something when the shaft reached 40 feet. It is not known if the shaft ever reached that depth, or if Oleson ever found any gold, but he did make the papers again in July. On Saturday the 18th, Oleson set off a blast which sent rocks flying through the windows of a neighboring house, striking the occupant, Mrs. Horne, in the face and wrist. The prospect was apparently shut down a short time later.

John M. Longyear

Gold mining fever in Marquette was revived in the late 1890s when the precious metal was found in copper deposits south of the city. The copper range, on what is now known as Migisy Bluff, was discovered in 1888 by Andrew S. Pings. Pings is best known for his brownstone quarry at Mount Mesnard. He was prospecting south of town for a company of local men when he found a vein of copper ore on the NE ¼ of the NW ¼ of Section 1, T47N-R25W. Along with T.C. and W.A. Foard, Peter F. Frei and Oliver Christmas, Pings worked for a while proving up the vein before selling the prospect to Julian Case of Marquette. Case was a mining man who had made his fortune in iron ore and was also involved in the Verde Antique marble quarry and several gold properties near Ishpeming and on the Gogebic range. He gave the new copper property his usual vigorous attention, sinking a shaft 18 feet into the vein and shipping a ton of the ore to a smelter in Chicago for testing. The ore, containing chalcocite and chalcopyrite rather than native copper, yielded more than 10½ percent copper, a very promising result. Unfortunately, Case died before he had a chance to develop the prospect.

After selling the prospect to Case, Andrew Pings continued to explore the Migisy Bluff area. He discovered

**J. M. LONGYEAR,**
BROKER IN
**Michigan Mineral and Timber Lands.**
MARQUETTE, MICH.

Agent for the Lands of the Lake Superior Ship Canal, Railway and Iron Company,
**600,000 ACRES FOR SALE OR LEASE.**

Lands Sold; Taxes Paid; Titles Examined; Defective Titles Adjusted; Lands Examined for Timber and Minerals; Lands Located at State and Government Offices.

**IRON ORE LANDS TO LEASE ON ROYALTY.**
Pine and Hardwood Timber for Sale.    References Given if Desired.

another copper deposit some distance west of the Case vein, near the center of Section 2, T47N-R25W. Pings was able to interest Peter R. Gottstein of Houghton, who obtained an option on the property from its owners, the Harlow family of Marquette, and hired Pings to head up the exploration. Gottstein and Pings sank a test pit and traced the vein onto the adjacent parcel of land, owned by the Iron Cliffs Company. When they were unable to get an option on the Iron Cliffs land, the prospectors abandoned their search, and the veins were forgotten for several years.

In the fall of 1897, explorations were resumed on the copper veins, spurred by reports that smelter tests of ore from the Case vein had yielded $12 in gold, 6 ounces of silver, and 3 percent copper. A wealthy Marquette man, John Munro Longyear, now owned the Case prospect, and set a small force of men to work sampling the vein. The Case shaft was located in the NE ¼ of the NW ¼ of Section 1, T47N-R25W. Longyear's crews traced the vein several hundred feet and sank a new shaft. A test of some 20 tons of ore at the Chicago and Aurora Smelting and Refining Company in Aurora, Illinois, showed less than 1 percent copper, however, and the prospect was closed down early in 1898. The shaft had reached a depth of 40 feet into the 10-foot-wide vein. (The Case prospect was explored again in the 20th century. Bear Creek Mining Company, the exploration subsidiary of Kennecott Minerals, drilled at least two diamond-drill holes into the Case vein in the 1980s with unpublicized results.)

Meanwhile, James Wilkinson, a Marquette banker, had obtained options on the former Gottstein and Pings vein from both the Harlows and the Iron Cliffs Company. Specimens from the Wilkinson prospect assayed as high as $10.33 in gold, with 1½ percent copper and a small amount of silver. In November 1897, Wilkinson sold his rights to the property to a group of Chicago investors

---

**IRON BAY M'F'G CO.**

MANUFACTURE

**Heavy Hoisting Machinery,**

**STEAM ENGINES,**

**BOILERS,**

**PUMPS AND PUMPING MACHINERY.**

MINING MACHINERY A SPECIALTY.

D. H. MERRITT, President and Treasurer.   C. Y. OSBURN, Secretary.

headed by John W. Ludwig, with Pings retaining a minor interest. The Ludwig group set four miners to work under the direction of Captain Josiah Broad, an experienced mining captain and former Marquette County sheriff, and sank two shafts east of Gottstein's test pit. One shaft on the Harlow property reached a depth of 40 feet, while the shaft on the Iron Cliffs land was sunk 25 feet. By January 1898, eight to 10 miners were working at the site, a boarding house had been built, and George Spencer had taken over as superintendent.

Early in 1898, Wallace Kirk, the son of a wealthy soap manufacturer from Chicago, came to Marquette and bought out the Ludwig company's interest in the prospect. Kirk increased the work force to 20 men and had an addition built onto the boarding house to accommodate them. It seemed that this infusion of money from the young millionaire was just what was needed to put the mine on a paying basis. When it came time to pay the bills, however, Kirk was always out of town, and when he finally did meet the payroll, his check bounced. When Kirk never returned, the miners all quit and the prospect closed again, this time for good.

Captain Martin Daniel may have seemed an unlikely prospector, being a Great Lakes ship captain rather than a mining captain. For more than 20 years he had sailed Lake Superior, often carrying cargos of black powder and high explosives from the powder works in Marquette to the copper mines of the Keweenaw Peninsula. As the master of the small schooner *Tom Boy*, he nearly lost his life to the lake when the heavily laden vessel sprang a leak during a storm and sank in August 1880. Captain Daniel and his lone crewman escaped in the ship's yawl and were picked up by a passing steamer. Within a month, the captain purchased a new vessel, the *Mystic*, and continued his hazardous voyages.

Some of his customers' enthusiasm for mineral exploration must have rubbed off on Daniel, for in the late 1890s, after retiring from the lakes, he began

---

**CAPT. M. DANIEL,**
SCHOONER "MYSTIC."
Dealer in
**LUMBER, LATH AND SHINGLES,**
AND GENERAL MERCHANDISE.
Goods of best quality, and at low prices. Coasting on Lake Superior a Specialty.
516 N. Front Street.

prospecting along the lake between Marquette and the Copper Country. In the winter of 1897-98, he discovered a vein of native silver about 18 miles up the shore from Marquette, apparently in the Saux Head region. That same winter, he also found a deposit of conglomerate rock containing 3 percent copper on the NE ¼ of Section 5, T49N-R26W. This is one of the only deposits of native copper ever found in Marquette County. Most of the other discoveries in this area have been copper ores, usually chalcopyrite or chalcocite. Along with Samuel York and John Tebo, Daniel started an exploration shaft, named the "Francis A" for his youngest child. By the end of 1898, a pool of Marquette businessmen had bought control of the prospect and set 10 men to work, deepening the shaft to 20 feet. The copper petered out, however, and the prospect was abandoned.

Captain Daniel, meanwhile, had directed his explorations a couple of miles to the north onto Section 30, T50N-R26W, where a homesteader had reported finding placer gold in a creek in 1891. Daniel soon found a four-foot-wide vein containing copper, silver and a trace of gold. Here he sank two test shafts, one at the bottom of a small hill and the other near the top. Assays showing up to 24 percent copper and $1.75 per ton in silver were rich enough to encourage Daniel to form a stock company in early 1899, and to offer 20,000 shares at 25 cents each. Treasurer George Barnes, a bank cashier, and secretary F. Stuart Byrne, a bookkeeper for the LS&I Railroad, handled the new company's finances. Revenues from the stock sale were used to hire miners from the Copper Country to deepen the No. 1 shaft at the base of the hill.

> **The Original Sauk's Head Mine Limited**
>
> Marquette County, Michigan

For the next three years, work was carried on sporadically as capital became available. Finally, on Christmas Eve 1901, the miners struck what they felt was the main body of ore at a depth of 30 feet, where the vein had widened to 12 feet. Although the copper content had decreased to 8 percent, the gold value was now $1.79 per

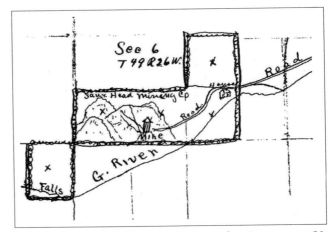

Frank Krieg's sketch of his family's gold mine.
FRED RYDHOLM

ton. With this promising showing, stock prices rose to 50 cents per share.

By the summer of 1902, the shaft was 50 feet deep and crosscutting was begun to test the width of the ore body. Copper now made up only 4 percent of the ore, but the gold was worth $3.75 and the silver $4.75 per ton of ore. The stockholders decided to reorganize the company as a development syndicate, issuing 1,000 shares at $100 par value. Peter Primeau and lawyer Charles F. Button joined Barnes as trustees, while George Hodgkins took over as secretary.

At 60 feet, the miners struck an even richer body of ore, which they concluded must be the true "main body." An assay by F.D. Tower of the Carp River Furnace showed 9.2 percent copper with $13.72 in gold and $3.99 in silver per ton. A steam hoist and water pump were added, and the shaft was deepened to 72 feet. Here the sinking stopped, while Daniel and his associates renewed their options on the property. In December 1902, the company was again reorganized, this time as The Original Sauk's Head Mine Ltd. The company offered 100,000 shares of stock for sale at 10 cents per share to finance further exploration and development work. Button served as chairman, Barnes continued as treasurer, and Alfred Archambeau succeeded Byrne as secretary. Louis Grabower and Joseph Vertin joined as managers.

Work resumed at the mine in February 1903, with eight miners working. Little news of the prospect was made public over the next year. Finally, in April 1904, the company bought the E ½ of the SE ¼ of the section, which the company's officers believed to carry the elusive "main vein," from the Rublein family. Apparently no main body of rich ore was ever found, and ownership of the land eventually reverted to the Rubleins. Today the shafts lie water-filled and forgotten, silent witnesses to Captain Daniel's dreams of golden riches.

After the "Francis A" prospect on Section 5 was abandoned in 1899, members of the Krieg family of Marquette began prospecting in the area. They traced the Daniel vein more than a mile to the west onto Section 6, T49N-R26W, where they owned 160 acres of land and held options on another 280. In November 1899, they formed the Saux Head Copper Mining and Development Company Ltd. Based in Detroit, the company was led by Captain John Considine Jr. of Detroit, president; Eugene J. Krieg of Marquette, vice president; John G. Krieg of Detroit, secretary; and George R. Purdon of Detroit, treasurer. Frank E. Krieg of Marquette served as superintendent. The company announced its plans to sell enough stock to begin development that winter, but work did not start in earnest until 1901, when an inclined shaft was sunk 14 feet before being shut down for the winter.

**LOUIS GRABOWER,**
**———THE LEADING———**
## Dry Goods and Carpet House
No. 309 S. Front Street,
MARQUETTE,   -   -   MICHIGAN.

Work was resumed in the spring of 1902, and the 8-foot-square shaft was deepened to 38 feet over the summer. The ore had richened steadily with depth, the gold value increasing from $4.16 per ton at 15 feet to $125 per ton at the bottom. In September, the company was reorganized and renamed the Saux Head Copper Mining Company Ltd. Detroit insurance agent Charles A. Stringer took over as president, with Charles Krieg as treasurer and Marquette lawyer Francis M. Moore

**Krieg's Gold Mine**
SUPERIOR VIEW

succeeding Eugene Krieg, who had died late in 1899, as vice president. The new company decided to equip the shaft with a steam hoist, to be housed in a log building along with the blacksmith shop. The hoisting engine and other equipment were brought in over the wagon road from Marquette, but the boiler was too large for the rock cut at Sugarloaf. Hauling the boiler over the rocks around the cut took several days. By the end of 1902, the shaft was down 54 feet, and a drift had been driven 12 feet into the vein. Assays from the shaft and drift showed gold values up to $195.70 and silver as high as $12.12 per ton.

In the spring of 1904, the Saux Head company hired expert mining engineer John A. Knight to supervise the prospect. Eight to 10 men were employed at the mine, and the shaft eventually reached a depth of about 120 feet. Assays of ore from near the bottom of the shaft in October 1904 showed $10.50 to $13 in gold and up to $75 in silver per ton, but water seeping into the workings from the nearby Big Garlic River finally forced the prospect to be abandoned that winter.

Frank Krieg stayed on at his homestead about half a mile from the mine after the shaft flooded. In 1905 he took a job running the Northern Lumber Company's general store at the newly built sawmill town of Birch, located between the mine and Saux Head Lake. Working with his brother Charles, he continued to prospect for gold on the family's lands. They sank two more small shafts, then decided to attempt to hit the ore body by driving an adit into the

Miners at Krieg's Gold Mine
SUPERIOR VIEW

hillside. The Kriegs drove the adit about 50 feet into the rock without finding gold before giving up for good.

Around 1891, after the closing of his mine near the Dead River, Ephraim Coon settled north of the small community of Clowry some 12 miles west of Ishpeming. Here, on the SE ¼ of Section 10, T48N-R29W, he started another shaft on a vein of gold-bearing quartz. Local legend says that a stranger named Hawkins mysteriously appeared and helped Coon to build a windmill on an 80-foot tripod to pump the water from the shaft. Just as the windmill neared completion, however, Hawkins disappeared as suddenly as he had arrived, leaving the machinery unfinished. Coon apparently never continued the mine.

In 1927, Martin Suneson bought 120 acres of land in the SW ¼ of Section 8, T46N-R29W, near the village of Republic. Over the next few years, he explored the land and diamond drilled, finding a vein of quartz 2 to 4 feet wide which he claimed carried up to $4 per ton in gold. The vein was too small and low grade to pay, however, and in the 1960s, when the Cleveland Cliffs Iron Company reopened the Republic iron mine as an open pit, Suneson's prospect was buried under a waste rock dump.

# CHAPTER 10

# Baraga, Iron & Dickinson County Prospects

During the silver lead era of the 1860s, most of the prospecting took place in the northern part of Marquette County, but some of the companies bought prospective mineral lands in what is now Baraga County. (At the time, Baraga and Keweenaw counties were part of Houghton County.) The Sioux, the Portage Lake and the Isabella Silver Mining companies, as well as John Forster's Fantine, Cosette, Marius, Valjean and LaPlata Mining companies, purchased land to the west of the Marquette County line hoping to find silver lead. None of these companies is known to have found a silver vein on its land. Marquette iron-mining pioneer Philo M. Everett bought 80 acres in Section 12, T50N-R30W during the silver boom, and his exploring crew reportedly found a 12-foot-wide vein carrying up to 35 ounces of gold per ton along with 7½ percent copper on his land, but the prospect was never developed.

In December 1863, at the beginning of the silver rush, a group of Detroit men, including Joseph Coulter, H.P. Sanger, W.A. Moore and Henry C. Kibbee, formed the Keweenaw Silver Mining Company. Coulter bought 240 acres in Section 20 T50N-R33W just east of Little Mountain and transferred it to the company. Why this land, which was more than 20 miles from the nearest known silver lead vein, was chosen for their explorations is a mystery, but may explain why no results were ever made public.

During the 1870s, a slate industry was starting up in Baraga County. Several quarries were opened on a formation of slate that ran for several miles east of L'Anse. The slate was found to be of high quality and suitable for roofing and other purposes. Around the same time, the Iron River silver mines in Ontonagon County were producing fine-grained native silver from similar beds of slate, so the Baraga County quarrymen were on the lookout for silver in their slate formations. In 1875, reports began to circulate that silver had been found in the L'Anse area.

Marquette Alderman Timothy T. Hurley had opened a brownstone quarry at L'Anse and also incorporated Hurley's Huron Mountain Slate and Mining Company to prospect for a bed of marketable slate. One of Hurley's slate prospects was rumored to have indications of silver and traces of gold, but no assay results were released. A Colonel Hodge of L'Anse found an outcrop of slate similar to the Iron River formation. An assay reportedly showed $12 per ton in silver, and the vein was confidently described as growing wider with depth, but its location was not made public. In December 1875, the Chippewa Mining Company was incorporated to work a silver vein near L'Anse, possibly the one discovered by Colonel Hodge. The company built a substantial boarding house and blacksmith shop and sank a shaft 30 feet into the vein, located on the E ½ of the SW ¼ of Section 6, T50N-R32W. Apparently the vein was also rich in graphite, since in February 1876 the Silver River Graphite Company was organized to work a vein of silver-bearing slate and graphite on the parcel.

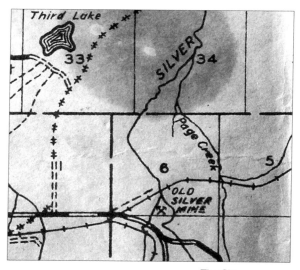

The Chippewa Mining Company's mine

ALL IMAGES AUTHOR'S COLLECTION UNLESS INDICATED

167

The 10-inch silver-bearing part of the vein assayed $34 per ton while the vein of graphite was about 26 inches wide.

During the fall of 1875, local newspapers reported that at least one of the prospects had been "salted" with silver. While the editors declined to overtly identify the property that had been salted, they did refer to it as being 20 miles from the head of Keweenaw Bay, which only applied to Hurley's prospect on the E ½ of the NW ¼ of Section 23, T51N-R30W. Regardless of whether they were legitimate or not, none of the properties was ever developed beyond the prospect stage.

Ever since the death of Douglass Houghton in 1845, rumors have persisted that his original gold find was located in the Huron Mountains of northern Marquette and Baraga counties. Even after the discovery of the Ishpeming Gold Range and the opening of the Ropes Gold Mine, prospectors continued to search this rugged northern country for gold veins. Late in 1884, reports began to be heard of a new gold mine somewhere between L'Anse and Michigamme, with rumors of ore worth $80 per ton. Finally, in May 1885, W.J. Ray of Ishpeming revealed the location of his prospect. Situated high on a divide in Section 17, T49N-R32W in Baraga County, the vein had been traced 500 feet along the surface. Five assays of the vein rock averaged $16 per ton in gold, and Ray announced plans to form a mining company, hire miners, and build a mill equipped with a Wiswell pulverizer. With the secret of its location revealed, public interest in the prospect dwindled, and little mention of it was made in the press. It is unknown if Ray's company ever came to be or how much work was done.

In 1882, before the discovery of the Michigan vein, George Grummett had found a gold-bearing vein on Section 20, T47N-R32W, in Baraga County. On the shores of what was then called Horricon Lake (later called Grummett Lake and now known as Ned Lake), he found the quartz vein in at least five spots. The lode ranged from 6 to 60 feet wide, with assays showing up to $30 per ton in gold. On the basis of this find, he organized the Grummett Gold and Silver Mining Company and purchased more than 260 acres, nearly the entire west side of the lake. Grummett served as president, with W.S. Hill as secretary and Anson B. Miner as treasurer. A force of miners were set to work under Captain Henry Davis in early 1885, but a lack of investors forced the company to close down the prospect. By that fall, Grummett had interested Detroit parties, and in the spring of 1886 he sold a one-half interest in the property to E.H. McGinley of Lower Michigan. How much work was done by McGinley is not known.

By 1906, gold mining was but a memory to most of Michigan, but the Grummett Gold and Silver Mining Company was still in existence and proposing to renew their explorations on the Michigamme prospect. At the company's 1906 stockholders meeting, a 2½-cents-per-share assessment was called to finance explorations. Over the summer, George Grummett and his sons returned to the prospect and reopened the three old shafts, which had been flooded since a logging company had dammed the outlet to the lake some years before. With the logging finished and the dam broken, Grummett was able to clean out the shafts, each of which was about 16 feet deep, and take samples of the ore in the vein. With specimens of free gold on display, the Grummett company offered the property for sale at a reasonable price, but no buyers came forward, and the prospect was again abandoned. In 1913 the property was sold to Randall P. Bronson, who had previously been involved in the B & M Gold Company, but no further work was done.

About seven miles east of the Grummett prospect, the Hancock Iron Mining Company was exploring for iron ore in 1888. One of the officers of the company was John R. Gordon of Hancock, who had found a gold vein on the Dead River Gold Range two years before. The Hancock company succeeded in finding a promising deposit of iron ore on the N ½ of Section 16, T47N-R31W, near Fence Lake. When the Michigan Gold Mine began to produce fabulously rich specimens that fall, the company's explorer, Captain John Marks, turned his attention to a vein of mineralized quartz that had been uncovered during the course of the exploration. Sinking a shaft through 25 feet of overburden to the vein, Captain Marks' crew found it to carry free gold. As the shaft was deepened, the vein grew wider and richer. Gold mining experts reported that the vein compared favorably with those in the Ishpeming area. The Hancock Iron company purchased a diamond drill rig and set to work looking for extensions of the vein, but must have failed in its endeavor, since no more was heard from this company.

Midway between the Grummett and Hancock prospects is what is rumored to be another gold mine. On the SE ¼ of the SE ¼ of Section 14, T47N-R32W, unknown parties drove a timbered adit some 50 feet into a gravel bank. The miners laid wooden rails and used a tram car to haul the ore (or perhaps gold-bearing gravel) out of the mine. Today the prospect is abandoned, the adit has caved in, and the tram car has been hauled away.

On some old maps of Baraga County, the notation "Old Gold Pit" is shown on the northern part of Section 10, T50N-R30W. Ownership records show that the NE ¼ of the section was purchased from the government by James Dwyer in 1885. Dwyer, who also owned a gold prospect near the Michigan Gold Mine, sold shares in the property to timber baron Timothy Nester, Henry Atkinson and Peter White. The property was later bought by mining engineer James Jopling and was eventually acquired by the Cleveland Cliffs Iron Company in 1901. CCI geologist C.M. Farnham, in a 1907 field report, described an exploration shaft sunk on a 6-inch quartz vein by the Marquette Gold and Silver Company in this area. It is not known if this company, which had worked

Terrence Moore's gold prospect in Marquette in 1883, actually explored this property, or if Dwyer and his associates operated under a similar title. In any case, the shallow shaft did not yield any quantity of gold.

In September 1888, a group of Copper Country men organized the Huron Mountain Exploring and Mining Company. The company was formed to explore and develop a vein of gold-bearing quartz that had been found by Thomas Dooling and Joseph C. Foley at an undisclosed location northeast of L'Anse in the Huron Mountains. Dooling served as president of the company, John F. Ryan was secretary and treasurer, and Foley, W.S. Cleaves and Martin Conway served as directors. The company was capitalized at $1 million: 40,000 shares at $25 par value. Information as to the location of the prospect and the extent of the company's work is lacking.

Also in 1888, John B. Belanger found a mineralized vein on the SW ¼ of Section 36, T52N-R30W, near the east branch of the Huron River. Belanger traced the 15-foot vein 100 feet on the surface and claimed that an assay of the rock showed $5 in gold, $10 in silver, and $10 in zinc per ton. Belanger never succeeded in opening a mine on the vein, but some time later the landowner, Christopher Ching, sold the parcel to John Thompson of Chicago as "valuable mineral land."

In 1914, Thompson was prospecting for manganese

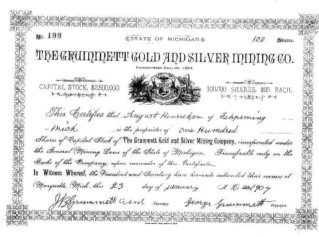

on his property and sent samples to Chicago for testing. Late in the year, the report came to him that some of the samples contained economic quantities of vanadium, which, like manganese, is used in making high strength steel, as well as a small percentage of uranium. Thompson put a crew to work at the prospect, building a snug winter camp and a blacksmith shop. By the end of the year, the men had sunk an 18-foot shaft, which Thompson's associate, George W. McGhee, named the Bughole Mine. They also started another shaft nearby, and although it only reached a depth of 4 feet, they optimistically called it the Mohatin Hills Mine. McGhee wrote a glowing report on the mine, which was published in *Mining and Engineering World*.

W.S. Cleaves

The article prompted R.C. Allen, the Michigan commissioner of mines and mineral statistics, to assign Professor W.E. Hopper of the Michigan College of Mines in Houghton to investigate the prospect. Hopper visited the site and took samples, but found the vein to contain nothing more than galena and small amounts of pyrite and chalcopyrite. The "mines" apparently closed down soon afterward, with much less fanfare than had accompanied their opening.

During the late 1940s and early 1950s, prospectors across the United States were looking for uranium to fuel the atomic power boom. The Jones and Laughlin Steel Corporation put summer employees to work surveying the company's lands in the U.P. with Geiger counters. In August 1949, those efforts paid off when the explorers found a vein of uranium on the east branch of the Huron River about a half-mile from Belanger's discovery. Detailed mineralogical studies of the vein in 2007 revealed an amazing array of minerals, including several vanadium compounds as well as silver. Perhaps Thompson and McGhee were right and Professor Hopper was wrong!

Orrin Robinson, an early mineral explorer, found gold at several locations in Baraga and Iron counties. Robinson felt that the numerous quartz veins crossing the rock formations near the upper reaches of the Sturgeon River were promising for gold. Around the turn of the century, he sank a 6-foot shaft on one of these veins and took out samples assaying from a trace to $2 per ton in gold. The vein proved too small to mine. Some years earlier, Jim Holliday, an Ojibwe man from L'Anse, had found a mineralized vein while trapping near Perch Lake in northern Iron County. He invited Robinson to inspect the vein. The men traveled by foot and canoe from Houghton to Perch Lake, where they relocated Holliday's vein near the north end of the lake. Assays of samples from the vein showed plumbago (graphite) and a trace of gold.

Orrin Robinson

Other locations in Baraga County were searched for gold in 1897 and 1898. Near Covington, an old explorer named Fred Schonnaway took an option on lands belonging to the Duluth, South Shore and Atlantic Railroad. He sank a 10-foot shaft, revealing rich looking rock, but no assays were made. At about the same time, a lumberman brought in samples of quartz carrying free gold worth $10 per ton. Other reports quoted assay reports as high as $12,000 per ton, but no locations were made public, and little more was heard of gold mining in Baraga County.

Dickinson County on the Menominee iron range was the site of numerous iron ore mines. In the late 1800s, it was also the site of several gold prospects. In 1881, a gold-bearing quartz vein in granite was discovered near Pine Creek about four miles north of Norway. Located on Section 34, T40N-R29W, the vein reportedly assayed $1,000 per ton. Land was bought and sold and options negotiated, but no amount of gold was mined. Four years later, during the summer of 1885, a small shaft was sunk some three miles northeast of Norway, apparently in the same area. One assay of a sample from the shaft showed $60 per ton in gold, while another showed $6 in gold plus $1 in silver. Apparently there was too little gold for mining, but enough to keep interest up, for the vein was still being tested more than 10 years later.

In 1888, state representative and veteran mineral explorer, Jonathan L. Buell, located a vein of gold-bearing rock on the banks of the Menominee River, south of the present town of East Kingsford. Buell was joined in the exploration of the prospect by Ben Marchand of Quinnesec and Joseph Flescheim of Menominee. Located on Lot 4, Section 7 and Lot 1, Section 8, T39N-R30W, the land was owned by the Menominee Mining Company and held under option by Hugh McLaughlin. Buell and his associates sank an inclined shaft, following the vein to an undisclosed depth. An assay of some of the vein rock by Julius Ropes showed $72 in gold and $10 in silver per ton, but the 6-inch vein was too small to work profitably. Today the steel rails of the skip road can still be found protruding from the filled-in shaft.

During the mid-1930s, two elderly men from Iron Mountain found a gold vein near the section line between Sections 19 and 30, T41N-R30W, about seven miles north of Iron Mountain. The quartz vein, which carried pyrite and chalcopyrite, also showed assay values from a trace to about an ounce of gold per ton. The men were unemployed and on relief, and were unable to raise the capital needed to work the prospect in the midst of the Great Depression, so the vein was abandoned. Years later the landowner filled in and leveled the prospector's test pits for the safety of his children.

Jonathan Buell

# CHAPTER 11

# Gogebic Range Gold & Silver Mines

Apart from the Ishpeming area, the only concentration of gold prospects in Michigan occurred on the eastern end of the Gogebic Iron Range. Before 1887, this roadless, undeveloped country was part of Ontonagon County. Several explorers were reported to have found gold and silver early in the area's history. A Dr. Kane and Mr. Purdy from Jackson, Michigan, as well as I.D. Bush of Detroit, were all said to have found the precious metals at undisclosed locations in the county. A Mr. McKellar (probably Peter McKellar, well-known geologist of Fort William, Ontario) reportedly found native gold in several townships in Range 46 West. None of these early finds were accurately located or developed.

The area between Lake Gogebic and Sunday Lake attracted many explorers looking for minerals, from iron ore to gold and silver. Lumberman and Wisconsin state Senator, John T. Kingston was one of those who took an interest in the area, and in 1879, he found a rock formation carrying argentiferous galena some four miles east of Lake Gogebic. He bought the property, the SW ¼ of the SW ¼ of Section 14 and the NE ¼ of the NE ¼ of Section 22, T47N-R43W in 1881 and sold part interest in the parcels to U.S. Representative Daniel Wells Jr. and Dr. George B. Miner of Milwaukee. The owners hired Captain Cornelius Gillis to give the land a thorough examination. Gillis found silver lead in several locations, but didn't immediately work toward developing the deposit.

John T. Kingston
ALL IMAGES AUTHOR'S COLLECTION UNLESS INDICATED

In the fall of 1883, Kingston, Miner and Wells organized the Lake Superior Silver and Lead Company to work the property. Captain Gillis continued to head up the prospecting crew and traced the vein across the property, finding it to be up to 75 feet wide and 800 feet long. Assays showed as much as $64.50 in silver, $9 in gold and $68 in lead per ton of ore.

Captain George Berringer took over from Gillis for the 1887 season, but work was not resumed after the winter shutdown. The prospect remained idle until late in 1888, when Edward Copps took over exploration on the vein. Copps and three associates took out a lease on the property in 1889 and formed the La Estrella Gold and Silver Mining Company to develop the prospect. The La Estrella company didn't seem to have any better fortune than their predecessors, as they surrendered their lease three years later without producing any lead, silver or gold.

Captain James Tobin was exploring about six miles west of Lake Gogebic early in 1884 when he found a vein of gold-bearing quartz in the N ½ of the NE ¼ of

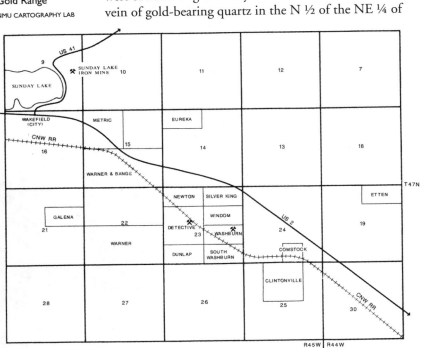

The Gogebic Gold Range
NMU CARTOGRAPHY LAB

Section 28, T47N-R43W, 1½ miles southwest of the Lake Superior Silver and Lead Company's prospect. The vein, which was 7 feet wide, carried gold in pyrite and copper ore and assayed $20 per ton. The Summit Exploration, Mining and Manufacturing Company, which was headed by George M. Wakefield of Oshkosh, Wisconsin, owned the land and hired Tobin to sink a shaft and test the vein. At a depth of 7 feet, the bottom of the shaft was sampled across its width, yielding an average of $30.65 in gold and $9.50 in silver. The shaft was carried down to 27 feet before the prospect was closed for the winter. Although the company announced plans to reopen the prospect in the spring, it seems the vein had pinched out at depth and no further work was done.

Peter Mitchell

Around 1880, Peter Mitchell of Ontonagon found a body of iron ore near the boundary between Sections 12 and 13, T46N-R42W a few miles southeast of Lake Gogebic. Mitchell sank a series of test pits, tracing the deposit along its east-west trend. The land belonged to the Agogebic Exploring, Mining and Manufacturing Company, which, like the Summit, was owned by investors led by George M. Wakefield. In February 1885, while exploring for iron ore at the Mitchell location, the Agogebic company's crews found a vein of gold-bearing quartz in their 85-foot-deep shaft. An assay of the rock showed $113.63 per ton in gold plus $12.10 per ton in silver. Although the vein proved too small for development, its discovery did prompt a boom in exploration of the adjoining sections.

The main beneficiaries of the boom seemed to be the Summit and Agogebic companies. The two companies owned 60,000 acres in the Gogebic area, much of it in the vicinity of the gold finds, and they did a good business selling exploration options. Near the Mitchell location, on the NW ¼ of the NW ¼ of Section 13, T46N-R42W, Julian Case of Marquette sank a 6-foot shaft into a quartz vein mineralized with copper ore. The vein produced samples assaying up to $182.81 in gold and silver. Captain Tobin and several other prospectors reported gold finds in Sections 12, 13, and 14, but none of these veins proved to be large enough for mining.

Today these prospects lie forgotten, marked only by abandoned trenches, shafts and adits. The mining history of the area is only recalled in names on the map, such as Gillis Creek, Prospectors Creek and Galena Creek.

Louis Vasseur was a pioneer settler of Ontonagon, having emigrated from Canada around 1853. He supported himself by fishing on Lake Superior, but also prospected for minerals when he could. Sometime before the Civil War, he discovered a silver-bearing formation near Sunday Lake in what later became Gogebic County. After serving in the Union Army during the war, he returned and explored the area where he had found the silver. In 1874 he and E. Hubbard of Oshkosh, Wisconsin, bought the S ½ of Section 10, T47N-R45W where the vein lay, then sold the property to Hubbard's associate, mineral and timber baron George M. Wakefield of Oshkosh. Wakefield set men to work investigating the silver vein in 1877, finding specimens worth up to $700 per ton in silver. No more work was done on the vein until 1883 when Congressman Richard Guenther, a partner with Wakefield in the Agogebic Exploring, Mining and Manufacturing Company, took an option on the property.

Louis Vasseur

Working for Guenther, Captain James Tobin sank two shafts on Section 10 in 1884 and traced the vein east and south onto the NW ¼ of Section 14, where he sank two more shafts. Assays of the vein rock showed values up to $215 per ton in silver. Later that year, a similar formation was found on Section 15. There must not have been enough of the silver-bearing rock to pursue mining it, since the property was later sold and became part of the Sunday Lake group of iron mines, which mined the soft red hematite ore which lay below the silver veins.

In the summer of 1885, Samuel Moreau found a 2-foot

vein of quartz carrying galena and pyrite about four miles southeast of Sunday Lake in the NE ¼ of Section 25, T47N-R45W. The vein assayed up to $15 per ton in gold and silver, but seemed to pinch out at a depth of 18 feet.

The iron-bearing formation that came to be known as the Gogebic Iron Range had been discovered as early as 1847 with the discovery of iron ore near Upson, Wisconsin. Over the next three decades ore was found progressively farther east, until the range was known to extend some 80 miles across Wisconsin and Michigan. It wasn't until the Milwaukee, Lake Superior and Western Railway was built through the area in 1884 that mining could begin. The Colby mine, which had been discovered years earlier, was the first mine to ship the rich ore of the Gogebic Range that year. With the railroad and the mines came an influx of people – prospectors to search for the next vein of iron, workers from the United States and Europe to mine the ore, merchants and businessmen to supply the needs of the booming iron range. Towns sprang up to house the newcomers – the village of Bessemer was platted by the railroad company near the Colby mine in 1884, and Ironwood grew up around the mines on the Michigan side of the Montreal River.

With the rapid growth of the Gogebic Range iron mines in the mid-1880s, George Wakefield saw the need for a new town near the west end of the range. The village of Wakefield was platted on the shores of Sunday Lake in 1886 and soon became a thriving community. Businessmen from Waupaca County, Wisconsin, were among the first to settle there. L.D. Goldberg of Marion, Wisconsin, opened a general store, selling groceries, clothing and miners' and lumbermen's supplies. His brother, B.M. Goldberg, set up a legal practice along with Walter S. Goodland, who went on to become the governor of Wisconsin in 1942. Goldberg and Goodland also published the new village's first newspaper, the *Wakefield Bulletin*. Land along the iron range was in great demand. Dozens of companies were formed to explore

George M. Wakefield

for iron ore, taking options on 80-acre parcels all along the presumed trend of the rich ore formation. A great many of these options were sold by W.W. Warner of Wakefield, a dealer in mining options and stocks.

In the spring of 1887, Wakefield hotelkeeper George Miracle was exploring for iron ore on the E ½ of the NE ¼ of Section 25, T47N-R45W on an option he held in partnership with L.D. Goldberg. Prospecting in the same quarter section where Sam Moreau had made his gold find, Miracle encountered a narrow seam of soft, decomposed rock that carried sulfide minerals. The seam grew wider as the exploratory shaft went deeper, and a sample was assayed, yielding $16 in silver and $6.60 in gold. The vein was traced some 275 feet along the surface, and a shaft was sunk 37 feet.

---

**W. W. WARNER & Co.**
—DEALERS IN—
**REAL ESTATE**
AND
**Mining Options.**

SILVER OPTIONS and STOCK
A SPECIALTY.

A Silver Formation Guaranteed to be LOCATED on Every Option Furnished by us, or Money Refunded.

All Legal Papers Executed with Neatness and Dispatch.
Notary Public        and        Collections.
WAKEFIELD,                                    MICH

---

With the promising showing of precious metals at the Miracle prospect, several exploration companies decided that, if their options didn't hold iron ore, perhaps they did have gold and silver. Parties from Clintonville, Wisconsin, including Dr. J. Finney, the town's mayor, organized the Clintonville Mining and Exploring Company in May 1887 to explore two 80-acre parcels just west of the Miracle vein. Other investors included Walter Goodland and B.M. Goldberg of Wakefield, as well as N. Etten, F. Nosack, J.A. Hickok and C.C. Spearbracker of Clintonville. Spearbracker served as president, Hickok was secretary, and Finney was the treasurer. The company's option included the E ½ of the NW ¼ and the W ½ of the NE ¼ of Section 25, T47N-R45W. The Clintonville company soon found a vein of silver ore which Goldberg and Goodland's *Wakefield Bulletin* touted as "large and valuable." Samples from the lode were sent to Chicago for analysis. Although the Clintonville management did not make public the results of these assays, they did proceed to sink a shaft. They announced plans to go to 65 feet, then start drifting, confident of striking the silver-bearing vein.

Mayor Finney and N. Etten were also officers in the Etten Mining Company. E. Brix of Clintonville was the president, with Etten serving as secretary and Finney as treasurer. The Etten company explored the N ½ of the NE ¼ of Section 19, T47N-R44W, a mile northeast of the Clintonville prospect without finding paying quantities of gold, silver or iron ore.

The most significant gold and silver strike on the Gogebic range came in the spring of 1887, when W.W. Warner found several small silver-bearing veins on the N ½ of the SE ¼ of Section 23. Having immediately secured an option on the property, Warner soon assigned his rights to the property to the Washburn Mining Company, which was formed by investors from Minneapolis and New York, including timber baron G.W. Washburn, president; A.D. Westby, secretary; and Captain D.F. Strobeck, treasurer. Warner retained a quarter interest in the property and joined with Strobeck, the company's mining captain, in prospecting the veins. The veins were found to come together into a central ore body, where Warner and Strobeck started a shaft. Assays of the galena- and pyrite-bearing rock from the shaft yielded from $8.49 to $138.48 in silver with a trace of gold. As the shaft went deeper, the assay reports became richer, and Captain Strobeck reported finding free gold in "sugar quartz" in the shaft.

Warner's mining option and real estate business grew rapidly as the gold fever took hold in the fall of 1887. Investors from Minnesota rushed to form mining companies and buy options anywhere near the gold and silver finds. Warner was ready to serve them, but, the local paper noted, "W.W. Warner cannot locate land situated on the gold and silver formation fast enough to supply companies formed to operate thereon."

The Detective Mining Company was formed by Minneapolis and St. Paul parties to work the S ½ of the NW ¼ and the N ½ of the SW ¼ of Section 23, west of the Washburn. Investors included W.M. Whitney of Chicago, president, C.S. Northrup of St. Paul, secretary, and

J.A. Northrup of Minneapolis, treasurer. The company claimed to have both the Clintonville and Washburn veins crossing their property, and cited assays of surface specimens showing $2.65 per ton in silver, with a trace of gold.

Minneapolis investors also organized the Dunlap Mining Company and took an option on the 80 acres adjoining the Detective property on the south, the S ½ of the SW ¼ of Section 23. Frank G. Turnquist served as president and John A. Turnquist was secretary and treasurer. The property showed a promising vein of quartz.

A.D. Westby, the secretary of the Washburn company, obtained an option on the N ½ of the NW ¼ of Section

---

THE
## DETECTIVE MINING
### COMPANY.

**CAPITAL STOCK $2,500,000 IN 100,000 SHARES.**

Situated on the S. ½ of N. W. ¼ & N. ½ of S. W. ¼ of Sec. 23, T. 47, R. 45 W.

### ✷ GOLD ✦ SILVER ✦ IRON ✷

W. M. WHITNEY, Chicago, Pres.   C. S. NORTHROP, St. Paul, Sec.

J. A. NORTHROP, Minneapolis, Treas.

**GENERAL OFFICE**

## WAKEFIELD, MICH.

BRANCH OFFICE

MINNEAPOLIS, - MINN.

SITUATED DIRECTLY ON THE SILVER AND GOLD FORMATION
WITH A WELL DEFINED OUTCROPPING OF A
RICH QUARTZ VEIN,
WHICH ASSAYS SHOW, CONTAINS LARGE QUANTITIES OF THE
PRECIOUS METALS, IT IS UNDOUBTEDLY A
WONDERFULLY RICH PROPERTY
AND OFFERS MAGNIFICIENT RETURNS FOR THOSE INVESTING.

23, north of the Detective. He assigned the option to the Newton Mining Company, which was owned by investors from Boston and Minneapolis, including William Leonard, president, L.O. Thayer, secretary, and O.S. Thayer, treasurer.

Just north of the Washburn property on the S ½ of the NE ¼ of Section 23, the Windom Gold and Silver Mining Company set men to work exploring. Along with the Detective, Dunlap, and Newton companies, the Windom's absentee owners left the operation of the prospect to Captain Strobeck.

The Windom prospect was bordered on the north by the property of the Silver King Mining Company. The company was formed by J.A. McCluskey, president; J.A. Northrup, secretary; and C.S. Northrup, treasurer. Operations were begun under the supervision of Captain Graham, an old western gold miner.

With this influx of capital to the region in the summer of 1887, George Miracle and L.D. Goldberg decided to form a stock company to work their Miracle prospect. They incorporated the Wakefield Gold and Silver Company and offered 1,000 shares of stock for sale. W.W. Warner was one of the first to purchase a large block of stock. Miracle was elected president of the new company, with Charles M. Harrington as vice president, Albert F. Olmstead as secretary, and Edward Copps as treasurer. In July 1887, Olmstead took samples to Ishpeming for assay by Julius Ropes. These specimens, from a depth of 28 feet in the shaft, showed $15 in gold. Other specimens were found to contain up to $100 per ton in silver. Over the next few months, the shaft was sunk to a depth of 60 feet, at which depth the lode was 16 to 36 inches wide. Arrangements were made to ship the ore from the mine to a smelter in Newport, Kentucky.

The prospects for silver and gold mining on the Gogebic range seemed promising enough in late 1887 that officials of the Washburn and Detective companies decided to buy a small smelting furnace to extract the metals from their ores. Shipping ore to the smelter in Kentucky for reduction cost $8 per ton plus transportation. By purchasing a Hartsfeld portable furnace for about $20,000, the mining companies could smelt the ore themselves at a cost of $4 per ton, Captain Strobeck claimed. The furnace seemed to be a fail-safe investment: if the gold and silver prospects didn't pan out, the furnace could supposedly be used to make pig iron from the local iron ore.

    Captain Strobeck, A.D. Westby of the Washburn Mining Company, and C.B. Holmes incorporated the South Washburn Mining and Smelting Company early in 1888 to buy and operate the smelter. The furnace was to be built on the S ½ of the SE ¼ of Section 23. The company also started an adit and a shaft into a bluff on their property and struck rock reported to be worth up to $60 per ton.

    Most of the gold and silver prospects were clustered southeast of Wakefield, but the formation apparently extended some seven miles farther to the west. Nick Bangs from Antigo, Wisconsin, took out an option on the E ½ of the SW ¼ of Section 15 T47N-T46W just south of Bessemer in 1887. Bangs discovered a vein of gold-bearing quartz on the parcel and sold his option to brothers A.F. and E.M. Nichols of Minneapolis. Major R.R. Henderson of Minneapolis joined the Nichols brothers in organizing the Old Veteran Mining Company, named in honor of Henderson's service in the Ohio Volunteer Infantry during the Civil War. The company started sinking a shaft on the vein and found ore carrying up to $60 in gold. They equipped the shaft with a steam hoist and sank it to at least 75 feet. A drift driven to from the bottom level struck a wide vein of quartz that assayed $12 to $40 in gold and silver.

    By the end of 1887, W.W. Warner had sunk the Washburn shaft to 40 feet and reported that the vein was still growing wider and richer. His own business was thriving, and his ads now proclaimed: "Silver Options and Stock a Specialty. A Silver Formation Guaranteed to be Located on Every Option Furnished by us, or Money Refunded."

    Some of the other prospects were not faring so well, however. At the Clintonville Mining and Exploring

> ## DUNLAP MINING COMPANY.
> # GOLD
> # SILVER
> # IRON
> ( SITUATION S½ of S. W. ¼ of SEC. 23, T. 47, R. 45, W. )
>
> **FRANK G. TWINQUIST, Minneapolis, Minn. Pres.**
>
> **JOHN A. TWINQUIST, Minneapolis, Minn. Sec. & Treas.**
>
> **GENERAL OFFICE, WAKEFIELD, MICH.**
>
> *Situated on the Slver and Gold Formation of the Gogebic Range, which is attracting the Attention of the Mining World, and which is developing rich veins and deposits of the precious Metals, it has a bright future.*
>
> ( THE PRODUCT OF THE MINE IS RICH IN MINERAL. )
>
> **A FINE OPENING FOR INVESTMENT.**

Company's prospect, the shaft had reached 65 feet, but pay rock was not to be found. A 20-foot drift to the west, likewise, failed to show up the elusive vein, and the investors gave up on the mine. A quarter mile to the east, the Wakefield company, too, seemed to be having trouble locating pay rock. The Dunlap company, which had been sinking on a vein of quartz, was still looking in vain for any strong indications of gold or silver. In October 1887, the prospect was sold to the Skandia Mining Company, which left it under the management of Captain Strobeck, but had no better luck finding riches. Except for initial assays of 47 cents to $13.19 per ton in silver from surface specimens, the Newton Mining Company never did report finding anything more valuable than "very promising quartz and greenstone."

The Detective and Silver King prospects were shut down over the winter. By March 1888, C.S. Northrup, who had been the secretary of the Detective and treasurer of the Silver King, was president of both and was trying to make arrangements to reopen the prospects, apparently without success.

W.W. Warner, ever the promoter, was not about to let the boom die down. He personally took over as manager of the Wakefield property and started sinking a second shaft 200 feet east of Miracle's shaft. He began to refer to the original 60-foot shaft as a "test pit" and confidently planned to sink the new shaft to 100 feet and then crosscut to determine the width of the vein. Over the summer of 1888, Warner's miners sank the new shaft at least 70 feet and drove drifts along the course of the vein. The new shaft was finely outfitted with a headframe and skip road. The Milwaukee, Lake Superior and Western Railway started a spur line into the Wakefield mine site in anticipation of deliveries of equipment to the property as well as shipments of riches from the mine. George Miracle and his fellow officers of the mining company, including his two sons, sold the property to W.W. Warner in the spring of 1889 when Warner's Washburn mine and mill were going strong. Despite Mr. Warner's grand plans to enlarge the Wakefield mine and install more machinery, the Wakefield Gold and Silver Company was never mentioned in the press again.

Warner also found new owners for part of the Clintonville prospect, selling it to the Comstock Gold and Silver Company. The Comstock company, optimistically named for the fabulously rich Comstock silver lode in Nevada, took over the eastern half of the Clintonville option and added the adjacent 40 acres to the north, the SW ¼ of the SE ¼ of Section 24. Warner managed the prospect for the company and declared that the vein in which the Clintonville shaft had been sunk was merely a feeder, and the "main vein" would be struck by drifting to the north. Later in 1888 the company started a new shaft north of the original shaft; assays from the new shaft showed $33.50 in silver with a trace of gold. A rich pocket was struck 50 feet down in the new shaft early in 1889, reportedly carrying "nearly free gold" in the quartz. Although the company drove an adit toward the "main vein" from the new shaft and announced plans to have a mill test done on its ore that spring, no more was news was heard from this prospect.

> # ✢ THE WASHBURN MINING COMPANY. ✢
>
> Situated on the N. ¼ of S. E. ¼ of Sec. 23, T. 47, R. 45 W,
>
> G. H. WASHBURN, New York City, Pres.   A. D. WESTBY, Minneapolis, Minn., Sec.
>
> D. F. STROBECK, Wakefield, Treas.
>
> ### GENERAL OFFICE,
> ## WAKEFIELD,   MICH.
>
> BRANCH OFFICE: Room 9, Chamber of Commerce, Minneapolis, Minn.
>
> # ❈ GOLD-SILVER-IRON ❈
>
> *A VEIN OF ORE BEARING ROCK*
> ## 9 -- FEET WIDE -- 9
> OF THE QUARTZ AND CARBONATE FORMATION.
>
> ## ❈ RICH ❈ IN ❈ SILVER ❈ AND ❈ GOLD. ❈
>
> A MAGNIFICENT DEPOSIT BEING OPENED UP.
>
> INFORMATION FURNISHED ON APPLICATION TO SECRETARY OR TREASURER.

The Windom prospect, too, was sold to new owners. Its Minneapolis owners had given it only a cursory exploration before selling it to another group of Twin Cities investors, who renamed it the North Washburn. Captain Strobeck retained his position as manager of the property.

By July 1888, the shaft on the Washburn Mining Company property had reached a depth of 60 feet, and a drift had been started to the east. Assays of the vein rock ranged from $3 to $28 and averaged $18 per ton. Apparently the South Washburn company never built their smelting furnace, for late in 1888 the Washburn company was reorganized as the Washburn Mining and Milling Company. The new company built a mill to crush and grind the ore and separate the silver and gold by amalgamation with mercury. The mill, which began operation in November 1888, was housed in a 45-by-90-foot, two-story building that also housed the boiler room, steam engine and an electric dynamo that powered electric lights in the mine and mill. The milling process employed a Wiswell pulverizer to finely grind the gold ore. Rather than amalgamating the gold and silver with mercury in the Wiswell machine as was the usual practice, the ground rock was fed into a pair of centrifugal pans where it was given a final grinding and mixed with several chemicals. After the mixture was heated by steam, it was fed into another pan where mercury was added to recover the gold.

---

# EXPLORERS!

**Wishing options for the purpose of**

## Exploring for Mineral

—IN—

## Marquette and Ontonagon Co.'s

Can receive them on application to the undersigned at Oshkosh, Wis. The lands are all well located on the mineral belt of the state, and include the lands belonging to the Tobin, Summit and Agogeebic Exploring, Mining and Manufacturing companies, in addition to my own, and other lands for which I am agent. Lists of lands furnished on application.     GEO. M. WAKEFIELD.
Oshkosh, Wis.

---

The mine and mill employed some 35 men, and ran day and night. Dr. N. Lehnen of St. Paul was the mining engineer.

Although many of the other prospects had failed, the apparent success of the Washburn kept investors interested in the potential fortunes to be made in precious metals in the Wakefield area. W.W. Warner located another gold- and silver-bearing vein on the NW ¼ of Section 15 in the summer of 1888. The Metric Gold and Silver Company was formed to operate the prospect. The company sank a 40-foot shaft on the vein, but it is not known what success they may have had.

Even though the Windom and North Washburn companies had failed to find pay rock on their land, W.W. Warner was able to interest a third company in a newly discovered vein on this property. Wisconsin's *Hurley Tribune* reported in January 1889, "At the Minneapolis and Gogebic Mining Company's property, adjoining the Washburn's on the north, a new and very rich vein of gold- and silver-bearing quartz has been shown up or discovered through the means of a new device or mineral rod, invented and controlled by Capt. W.W. Warner, manager of the Comstock mine, who has located all the veins and shafts of the Washburn properties. This vein on the Minneapolis property was located by the instrument at night, and on doing some stripping next day the quartz vein was discovered to be literally full of galena and silver ore. A fair sample of the quartz was assayed and ran more than $50 in silver, with traces of gold. It will pay mining men to have their properties examined by Mr. Warner before sinking when there are no indications on surface, as the instrument used is pretty certain to give the indications of mineral if any exist."

The Washburn mill was able to treat only 12 tons of rock per day, being limited by the small capacity of the Wiswell pulverizer, but the company confidently planned to add 100 heads of Cornish stamps, an air compressor and power drills, and to sink the shaft to 400 feet with drifts every 50 feet. These plans were never carried out, since the company had spent all of its investors' capital and was nearly $5,000 in debt. Their main creditor, the Bank of Wakefield, had the company's assets seized and sold at a sheriff's sale in June 1889. This spelled the end for the Washburn Mining and Milling Company, as well as for the stock-promoting career of its founder, W.W. Warner, who seems to have disappeared from the Gogebic Range around this time.

The Washburn never produced any appreciable amount of bullion. The complicated process of extracting the gold was apparently too expensive, and the vein was too narrow to be worked at a profit despite all the rosy pictures painted by the mine's promoters. The shaft only reached a depth of 100 feet, with a single 40-foot drift. Drifting and crosscutting in the South Washburn adit totaled only 120 feet. With its grand size, nearly 8 feet high and 10 feet wide with perfectly squared-up sides and ceiling, the adit undoubtedly impressed visiting investors, but apparently never produced any significant amount of gold or silver ore.

A 7-foot vein of rich gold-bearing quartz had been struck west of the shaft at the Old Veteran prospect south of Bessemer in 1889. In the fall the company decided to sink a new shaft and to equip the mine with steam drills. The new shaft intersected the old drift early in 1890, but the gold-bearing vein must not have panned out, as the company was never mentioned again.

W.W. Warner had also participated in several other explorations. He did some prospecting on the W ½ of the S ½ of Section 15 adjacent to the Old Veteran, and on the S ½ of Section 22, reporting galena with silver on both locations. He also explored on Section 21, and was reported to have sunk two shafts to 15 and 35 feet. A resident of the SW ¼ of the NE ¼ of this section presently uses an old shaft as his well, which he reports to be 80 to 90 feet deep. The Galena Mining Company had explored on the S ½ of the NE ¼ of the section and reported finding silver, lead and iron.

Late in the 19th century, George Triplett and his wife farmed and raised a family of eight children about 10 miles north of Ironwood. Their homestead was on Section 6, T48N-R47W on the east side of what is now called Triplett Road. Around 1911, Triplett started prospecting in the hills north of the farm, looking for precious metals. Despite geologists' opinions that he was looking in the wrong kind of rock formation, he found indications of copper and silver. Triplett continued to dig for years, sinking test shafts and driving adits without ever finding his fortune.

Other companies which prospected for gold on the Gogebic range in Michigan included the Eureka Company, located on the N ½ of the NW ¼ of Section 14, T47N-R45W, the Jay Gould Mining Company on Section 15, T47N-R45W and the Section Twenty Five Company. Little is known about these other companies, but they are all part of the story of the short-lived gold and silver rush on the Gogebic range.

## Chapter 12

# Placer Mining

In prospecting for gold, the '49ers of the California gold rush would wash gravel from the stream beds in pans and sluices to extract the metal deposited there by erosion. As these placer deposits played out, the prospectors followed the rivers upstream, looking for the "mother lode," the gold-bearing bedrock from which the secondary deposits had been eroded.

Mineral explorers in the Upper Peninsula panned the many creeks and rivers, and occasionally found some "colors," but no bedrock deposits were ever known to be located in this way.

The geography of the U.P. differs from that of California in that the earth's surface in northern Michigan has been scoured and reshaped by the repeated advance and retreat of continental glaciers during the Pleistocene Epoch. The sand and gravel (and gold) in a stream bed here could have been eroded from bedrock nearby, or it could just as easily have come from hundreds of miles away, carried by an advancing glacier. Thus, gold in a river placer was no proof that a

William M. Courtis
ALL IMAGES AUTHOR'S COLLECTION UNLESS INDICATED

mother lode lay upstream. Nonetheless, gold placers are known to exist in Michigan, and mining them has often been proposed.

In the Report of the State Board of Geological Survey of Michigan for the Year 1906, William M. Courtis of Detroit, a mining engineer formerly associated with the Peninsular Gold Mining Company, listed the following placer gold finds:

> Allegan County: Allegan.
> Antrim County
> Charlevoix County: Boyne River.
> Emmet County: Little Traverse River.
> Ionia County: Maple River, Grand River below Lyons.
> Iron County: Iron River.
> Kalkaska County: Kalkaska, Rapid River, Walton.
> Kent County: Lowell, Ada Creek.
> Leelanau County: near Lake.
> Marquette County: Ishpeming.
> Manistee County: Little Sable River, Manistee River.
> Montcalm County: Greenville, Howard City.
> Newaygo County: County Line, Muskegon River.
> Oakland County: Birmingham.
> Oceana County: Elbridge, Hart, White River, Whitehall.
> Ontonagon County: Flat River, Victoria copper mine.
> Ottawa County: Grand Haven.
> St. Joseph County: Marcellus, Burr Oak.
> Wexford County: Manistee River, West Summit.

Although a number of these finds were authenticated, others were believed to be just pyrite. It is doubtful that any of them are rich enough to be worked profitably.

The Carp River, which runs through the Ishpeming Gold Range as well as passing through large deposits of glacial drift, has been the source of a few placer gold reports. Around 1851, William A. Burt is reported to have found a nugget of gold in the Carp while building a dam. The nugget, which was the size of a bean, apparently was the only gold found. It didn't inspire any

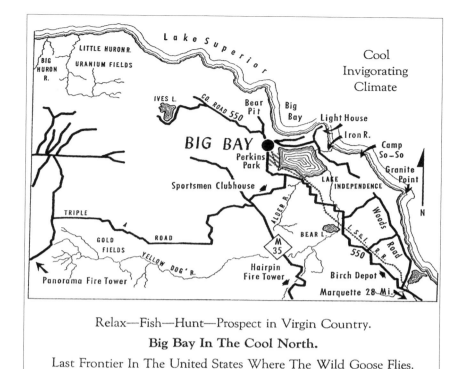

AL WOOD

large scale prospecting. Some years later, geologist Forest Sheppard was also reported to have found placer gold in the river.

In a report on the Dead River Gold Range written about 1890, Julius Ropes proposed placer mining in the many valleys in this rugged area. "That placer gold exists in quantities over all this area, to be profitably panned, cradled or sluiced, there is no doubt, tests made of sands and gravel in the ravines give every assurance," he wrote. Ropes was never known to do any placer mining in the region, however.

Some seven years before Captain Daniel opened the Original Sauk's Head Mine on Section 30, T50N-R26W, placer gold had been found in the same section. In a letter to the Marquette *Mining Journal*, homesteader George Rasmussen reported that he had taken samples of the sand from "a small stream of the finest running water you ever tasted" (apparently Sawmill Creek) as well as samples of rock from the area. When he brought them to Marquette for assay, the chemist reported both the sand and the rock samples showed small quantities of gold. Rasmussen later moved away from Michigan and never resumed his prospecting.

The Yellow Dog Plains, an area of glacial sand and gravel deposits in northern Marquette County, have long been thought to contain placer gold. One of the earliest settlers on the plains, Nels Andersen, homesteaded on the SW ¼ of the SW ¼ of Section 35, T51N-R29W from 1902 to 1913. While digging a well, he encountered a 2-foot layer of black sand at a depth of 45 feet. Andersen felt that the black sand held gold and, some years later, had a sample assayed. His son claimed the assay showed $5.35 per ton in gold. While prospecting for the Michigan College of Mining and Technology in 1933, T.M. Broderick took a number of samples of sand and gravel from the Yellow Dog Plains and tested them for gold. A sample taken from Andersen's diggings proved to be the richest, showing 20 cents in gold per ton. Other samples from the plains assayed from 3 cents to 16 cents per ton.

An area of the Yellow Dog Plains southeast of Andersen's homestead was prospected in 1932 with dubious results. The McKay Company of Pittsburgh set crews to work digging test pits and taking samples at numerous locations on the plains. A total of 107 samples were asssayed, with values ranging from 12 cents to $2.69 per ton. The McKay Company's officials did not feel that the showing was worth developing and left the area before completing their negotiations with the owner of the land.

In 1934, the state geologist, apparently encouraged by Broderick's findings, proposed to search along the Yellow Dog River for placer gold. An agreement was reached whereby the landowners would furnish the necessary equipment, while men from the Civilian Conservation Corps camp at Big Bay would provide the manpower. The proposal was rejected by the Michigan Conservation Commission because it didn't fall within the scope of the Corps' work.

During the Great Depression, a few men worked the sands of the Yellow Dog River and other streams that cut through the glacial outwash plains. Working on their own, these rugged individuals were sometimes able to eke out a living by panning and sluicing for gold. One man claimed to have averaged $5 per day while working a small stream in the northern part of the county. Even years later he refused to reveal the name of his bonanza creek.

The Yellow Dog Plains were sampled again in the 1970s by two mining companies and the U.S. Geological Survey. The results were disappointing to say the least: One of the companies reported finding only one "color" in 51 samples. The USGS reported finding no "colors" in 32 samples, and their assays showed less than .0006 ounces of gold per ton.

In the late 1960s, retired electrical engineer Burnell Tindall and Don Malott, a Michigan State Highway Department expert on sand and gravel, along with two other partners, formed Au Min Company Inc. The company obtained contracts allowing them to work gravel pits in Clinton, Oakland and Washtenaw counties in the Lower Peninsula, sluicing the sand and gravel for heavy minerals and extracting gold from the concentrate by a secret process. Although at one point Tindall boasted that they had recovered $2 million in gold, actual recovery was much less. During one year, the company extracted $101 in gold, but incurred expenses of $20,000. When the other three partners moved away or died, Tindall kept up the work. Although yields increased, all work ceased with his death in 1976.

# Chapter 13

## Fables & Frauds

Many of the gold discoveries in Michigan's Upper Peninsula were too small or too poor to pay, but they generally held some amount of real gold. Along with these bona fide discoveries, however, the state also had its share of mythical mines, both fabulous and fraudulent.

Beginning in 1872, the Iron River area of Ontonagon County near present day Silver City was the site of a silver mining boom. Veins of rock carrying native silver were discovered, and mining companies were formed to buy the land and work the veins. Several assays of the silver ore showed traces of gold, but the nature of the vein made the assayer and other geological experts doubt that gold would be found there. Eighteen months later, it was discovered that the specimens in question had indeed been salted with gold. Although the silver mines continued to produce in a small way for a number of years, no more talk was heard of gold mines in the area.

Like any mining area, Upper Michigan is home to many stories of "lost mines," tales of rich veins of gold and silver found by accident, never to be seen again. "Larsen's Gold" is a typical example of these tales. In *Bloodstoppers and Bearwalkers, Folk Traditions of the Upper Peninsula*, author Richard M. Dorson quotes Jim Hodge of Negaunee:

"A fellow named Larsen was travelling north of Teal Lake, Negaunee, picking berries, lots of berries, on the hills. He picked up a piece of quartz, attracted by the specks in it. If he'd been a mining man, he would have known there was gold in it. He put it in his pocket and cut across the hills to town to his brother's saloon. There was a mining man present; he said, 'Show it to me.' After he looked at it, he said, 'I'd like to show that to a friend of mine.' He sent the piece to an assayer at Houghton, and it was assayed at $200 a ton in gold. But when Larsen went back to look for the quartz, he couldn't find the spot, or even the hill."

Perhaps the legend had a basis in fact, considering the gold find at the Korten prospect north of Teal Lake, but the value of the rock grew to legendary proportion.

Dorson also tells the tale of an elusive mine in the western Upper Peninsula, quoting Aaron Kinney of Iron River: "Peter Paul's Gold Mine is on the south branch of the Paint, right below Uno dam. Peter Paul came from Canada, where they had gold, and here he found a piece of quartz that looked like gold. He was telling my father he'd strike the gold in a few strokes. He blast drilled, sunk a shaft, single jacking, twisting the drill himself. When he came down to water he had to quit, because he had only a hand pump. He was going to sell blueberries on the land around there to make a stake to develop his gold mine. People could pay him so much a quart to pick themselves. There were blueberries a solid mile up the river, you see. The dream of the gold mine never left him, but he never got a stake. He dug until he was too old to dig and died without making a fortune. His place is still called Peter Paul's Gold Mine."

Peter Paul's mine can still be seen today on the east bank of the Paint River, a few hundred yards off the Goldmine Truck Trail (USFS 151) between Beechwood and Gibbs City. Located on the SW ¼ of the NW ¼ of Section 13, T44N-R36W, the mine site includes the remains of the shaft and rock dump, as well as the ruins of Paul's homestead.

Around 1885, a man named Jack Grove crossed illegally from Canada into the Upper Peninsula and took up residence in the wilderness north of Champion. Grove had been a prospector in Canada. When he returned from one prospecting trip minus his partner, he was sought for questioning by the Royal Canadian Mounted

Police. Rather than face the Mounties, he decided to go to the United States, where he lived alone for the rest of his life.

While living his solitary life in a series of shacks near the headwaters of the Huron, Peshekee and Yellow Dog rivers, Jack Grove continued his prospecting and claimed to have found veins of gold-bearing quartz in several places near his camps. He wrote to the landowner, William C. Weber of Detroit, asking him for a one-half interest in any gold he might discover. Weber refused to deal with Grove, but sent his own men into the woods to prospect his lands. Weber's men never found any gold, and Grove died without ever opening a mine or revealing his secret.

Although traces of gold can be found in glacial gravels throughout Michigan, no economic deposits have ever been found in the Lower Peninsula. Nonetheless, the downstate area was host to a number of cases of gold fever. One of the earliest examples occurred in January 1848, when it was reported in *Hunt's Merchant's Magazine* that a gold strike had been made near downstate Tecumseh. Nuggets weighing up to an ounce and a half were rumored, and a company was to be formed to work the deposit. Apparently little work was done before the strike was found to be worthless.

The Alpena area was the site of one of the most successful "pocket mines" in Michigan (so named, not for the geological nature of the deposits, but because the only gold mined was from the pockets of investors).

In 1924, promoter from Chicago formed a company to mine for gold in the limestone of the region. Although there was no gold visible in the rock and independent assays failed to find any gold, the promoters assured potential investors that the "ore" held extremely finely divided gold that was too fine to be detected by normal assaying methods. The company's chemist laid claim to a wonderful method of recovering the fine gold. Investors from Michigan and Chicago were persuaded to put money into the venture, and the company diamond drilled in search of more gold, and leased part of Angus Morris' farm

LEE DEGOOD

north of the village of Herron. Here they sank a shaft 230 feet at a cost of $200,000, and a mill to put the chemist's method to work was built at a similar cost.

Within about a year, the Alpena gold boom died out. Investigations by the Michigan Geological Survey revealed that ore samples had been salted. In one case, a specimen had been drilled out and gold inserted into the hole. In another, a sample had been treated with a chemical solution that deposited gold in the rock. With these revelations and with the investors' capital spent, the company's officials quickly and quietly left town.

The pocket mining business was revived around 1930 with the organization of the Thunder Bay Gold Corporation, which issued handsome stock certificates, sold shares and reopened the mine for a short time before going bankrupt, leaving the empty shaft and silent mill as monuments to the gold seekers' eternal optimism.

With the 70 percent increase in the price of gold in 1934, gold fever again took hold in the Lower Peninsula, and even in Alpena again. Once again, a mining engineer discovered "micronic" gold, undetectable by ordinary means, and again a chemist came up with a means of extracting this elusive gold. This time, the charlatans were quickly unmasked and were forced to find other sheep to fleece.

In 1952, two men from Chicago decided that the old gold mine on the Morris property would be a good place to prospect for uranium. Along with four local men, they sank the shaft deeper and sampled for uranium. While drilling for some final rock samples before shutting down for the winter, a spark from the miners' tools touched off a pocket of gas that had accumulated in the mine. The resulting explosion killed the two men underground and three more at the mouth of the shaft on surface. No gold, no uranium, but five men dead in their futile search for riches.

A gold "rush" at Vernon in Shiawassee County was typical of the gold excitement of the 1930s. An article in the *The New York Times* told the story of an Ojibwe man who had made his living panning gold from the Shiawassee River many years before. In 1935, a grandson of the old Ojibwe man showed up with a birch-bark map pinpointing the location of the gold find. The grandson and his associates followed the map and started digging. The paper reported that the deposits turned up showed assay values equal to the best Klondike finds. Guards were stationed at the diggings, and the ore was sent to a smelter in Detroit.

Similar gold booms occurred throughout the Lower Peninsula in the mid-1930s, notably in Montrose, Ortonville, Perry and Grand Rapids. It seems that none of these "pocket mines" ever panned out, but made money only for their promoters.

Even the Ropes Gold Mine, where gold mining had been proven to be economically feasible, was the source of at least one picturesque lie. In its October 31, 1894, issue, the *Chicago Inter Ocean* ran this "Special Telegram":

"A report comes from the Ropes Gold Mine, near here, that at the bottom level a spring has been struck which supplies a strong stream of highly colored water, being nearly as yellow as gold and plainly holding considerable of that precious material in its solution. It is very palatable and ice cold. Several of the men working in that level have been in the habit of freely drinking the water and it was noticed that as they drank the desire for alcoholic stimulants died within them. Some of the men have been heavy drinkers of beer and whiskey but since the use of this water they had no use for any stimulants. Their health and physical condition is generally improved, and it is thought that the water is a veritable and natural 'gold cure,' not only for the liquor habit but for general diseases of all kinds. There is some talk of erecting a huge hospital at the mine, using the water as a cure-all for the ills of men."

The story was quickly picked up and debunked by the local papers (such a solution of gold in water being chemically impossible) but not before the mine management received a number of inquiries about the "gold cure."

# CHAPTER 14

# Epilogue - 20th & 21st Century Explorations

Although none of the prospects discovered in the 19th and early 20th centuries ever turned into profitable mines, the Ishpeming Gold Range continued to attract prospectors through the 20th century and into the 21st. Exploration and mining companies prospected the old workings and searched for new veins, with the same varying results as the old-timers.

Among the companies that participated in this continuing search for precious metals was the Cleveland Cliffs Iron Company, working both independently and in joint ventures with others. In 1966, Cleveland Cliffs diamond drilled at the Michigan Gold Mine, looking for continuations of the gold veins. Cleveland Cliffs and Bethlehem Steel formed the Beth Cliffs Joint Venture and explored for precious metals in the 1970s. In 1977, Cliffs joined with Chevron Oil as the Nomex Au Joint Venture and diamond drilled at Bjork and Lundin's exploration and other locations west of the Ropes Gold Mine, as well as exploring for copper in the Kona formation south of Marquette. None of Cliffs' non-ferrous metal explorations paid off, however.

Between 1968 and 1971, the Humble Oil Company (later to become Exxon) explored for massive sulphides and precious metals north of the Dead River and diamond drilled near Clark Creek and in the Reany Lake area. A joint venture involving the Superior Oil Company and Nicor Mineral ventures also prospected in the Dead River area from 1972 to 1985. This joint venture drilled 14 diamond-drill holes in the vicinity of the Holyoke Silver Mine. Other companies known to have prospected on the gold range include Bear Creek, Kerr McGee, Phelps Dodge and St. Joe American Corporation.

Systematic geological mapping of the greenstone belt north of the Dead River was undertaken by the Geological Survey Division of the Michigan Department of Natural Resources and the Department of Geology and Geological Engineering of Michigan Technological University between 1984 and 1989. The project involved Michigan Tech graduate students and faculty as well as DNR geologists. The geology of more than 40 square miles of the greenstone belt was mapped and sampled, and several areas of anomalous gold and silver concentration were identified. This work provided a baseline for future exploration work by mining companies.

After the closing of the Ropes Mine, Callahan Mining Corporation continued its exploration program, hoping to find further reserves of gold-bearing rock. The company held mineral rights to the Michigan, Superior and Peninsular gold mining properties west of the Ropes, where diamond drilling was done to evaluate the ore bodies. Ore found at the old Peninsular gold mine was rich enough that at one point Callahan considered reopening the mine. The company also did diamond drilling and intensive sampling at Bjork and Lundin's exploration west of the Ropes and in the Silver Creek area north of the Dead River. Early in 1991, Callahan entered into a joint venture with Western Mining Corporation (USA) and Cleveland Cliffs Iron Company to explore for gold in the vicinity of the closed Champion iron ore mine, some 13 miles west of the Ropes. None of these explorations matured into mining ventures, and Callahan sold its mineral rights to Minerals Processing Corporation when it left the area, ending its involvement in Michigan gold and silver mining.

The end of the 20th century seemed to usher in a new era of precious mineral exploration, in sharp contrast to the old days. The popular stereotype of the grizzled prospector with his pick and shovel is not too far off when it comes to early mineral exploration in the Upper Peninsula. Many mineral discoveries were made by individuals or small groups of independent explorers. Most of the land, apart from developed areas, was owned by the federal government, so mineral explorers roamed the country freely, looking for valuable ores. Although many of them lacked formal schooling in geology and mineralogy, practical experience and lore handed down from other explorers often served them well.

Prospectors for precious metals in the Upper Peninsula knew to look for intrusive formations such as quartz veins in igneous and metamorphic rock. Within the quartz veins, indicator minerals, including metallic sulfides such as pyrite and galena, would often give clues to the presence of gold and silver. Once a promising vein had been located, it would be tested for the precious metals. At its simplest, this could be done by pounding the rock to a powder with a mortar and pestle (or simply an iron bar and a heavy cast iron cook pot) and panning it out in a gold pan. For a more precise test and to detect finely disseminated gold and silver in sulfide ores, a fire assay was required. This physical and chemical test could accurately detect and measure the amount of various metals in the sample.

With the precious metal content of a vein determined, the only method the prospector had to measure its extent was to sink a shaft, following the vein into the earth to see if its size and richness persisted. In the latter part of the 19th century, the diamond core drill came into use in the iron ranges of Michigan and was soon put to use in gold and silver exploration. A diamond drill uses a tubular drill rod turning a bit faced with industrial diamonds to cut out a cylindrical core of the rock it is drilling through. The drill is rotated at high speed while a stream of water is pumped down the hollow center of the drill pipe to flush out the rock chips produced. Periodically the drill pipe and bit are withdrawn from the hole, bringing the core of rock with it. This core is examined and assayed to determine the composition of the bedrock underground.

The diamond drill removed the need to sink shafts to test prospective ore veins, but the high cost of the machinery meant that only larger mining companies could afford to use the technology. Small operators continued to sink shafts to explore their finds, drilling holes by hand and blasting with black powder or the new "giant powder" (dynamite). The blasted rock was loaded by hand into an ore bucket or kibble and raised to the surface by a hand winch or horse whim. Only when a mine had been developed into a producing mine was power equipment such as power drills and steam hoists put into service.

Exploring for precious metals was largely unchanged for the first part of the 20th century. Prospectors continued to look for promising rock formations exposed at the surface. Nearly all of the economical mineral deposits that cropped out on the surface had been discovered and explored, but valuable deposits undoubtedly lay hidden beneath many feet of soil and rock. The development of geophysical exploration technologies later in the century opened up a new era in mineral exploration. A number of techniques were developed, which enabled geologists to infer what lay hidden below the surface.

The force of gravity at the earth's surface is dependent on the distance from the center of the earth and on the mass of the earth below. Sensitive gravity meters allow geologists to measure gravity very precisely. By correcting for elevation and geographic factors, these gravity readings can indicate the density of rock formations up to several thousand feet below the surface. High-density rock can be an indication of massive sulfide deposits or other metallic ore-bearing rock formations.

The earth's magnetic field at the surface varies according to the composition and structure of the rock underground. Early prospectors in the U.P. compared the readings of a magnetic compass with a solar compass to detect bodies of magnetic iron ore. They also used a variation of the compass called a dip needle, which had a magnetized needle that pivoted vertically to measure the inclination of the magnetic field and delineate iron-ore formations. Modern electronic magnetometers have improved on the compass and dip needle and can precisely measure the magnetic field in three axes.

Modern electromagnetic exploration methods are able to detect conductive mineral formations underground by measuring currents in the earth that have been induced by varying magnetic fields. Some implementations of the technology send out electrical pulses, which induce a magnetic field in the earth's crust, then measure the returning signal that is picked up by a sensitive receiver, much like a beachcomber's metal detector. Other technologies rely on natural phenomena, such as thunderstorms to induce the magnetic fields in the crust, while others use existing very low-frequency transmitters hundreds to thousands of miles away. These technologies are able to detect conductive bodies such as metallic ore deposits.

Despite the availability of all these geophysical technologies, valuable mineral deposits can still be found by careful footwork, observation or just plain luck, as was proven by the discovery of the Back Forty zinc, gold, silver and copper deposit. In 2001, a well driller was sinking a well for a landowner along the Menominee River in western Menominee County. The last 35 feet of the well passed through an odd sparkly rock that the driller recognized as a metallic sulfide. The well driller showed the drill cuttings to a geologist, who analyzed the ore and identified the mineral as sphalerite, an important ore of zinc, and confirmed that it carried 10 percent zinc. The geologist asked the landowner to show him all of the bedrock exposures in the area, and within 200 yards of the well, they found an outcrop of rusty-looking rock. This turned out to be gossan, an oxidized form of massive sulfide, and carried up to an ounce of gold per ton.

The well driller and the geologist partnered with exploration company Minerals Processing Corporation, and the partnership quietly bought and leased as much of the mineral rights to the surrounding land as they could. Gravity and electromagnetic studies of the area gave strong responses, leading to a diamond-drilling program that discovered a major body of rich sulfide ore carrying zinc, gold, silver and copper in spring 2002. Minerals Processing Corporation and its partners formed a publicly held company, Aquila Resources, to continue exploration and to develop a mine on the deposit.

Over the next 11 years, Aquila and partners drilled more than 500 diamond-drill holes totaling nearly 80 miles, some as deep as 2,300 feet. The drilling delineated an ore body known as a volcanogenic massive sulfide (VMS) deposit, which had been formed by hot mineral springs (black smokers) in volcanic rocks at the bottom of a prehistoric ocean. The Back Forty deposit was formed 1.87 billion years ago, when what is now the Upper Peninsula was part of an oceanic volcanic island arc where the earth's crust was being pulled apart and molten rock was welling up toward the surface and forming underwater volcanoes. Springs of superheated water associated with these volcanoes brought dissolved sulfur and metals such as zinc, copper, lead, gold and silver into the ocean waters, where they cooled and deposited sulfide minerals in beds on the sea floor. Once the volcanoes cooled and the black smokers stopped flowing, the sulfide deposits were buried by other volcanic rock and marine sediments that eventually hardened into rock.

In August 2009, Aquila entered into an agreement with HudBay Minerals Inc., a Canadian mining company. For an infusion of capital, HudBay acquired 14.9 percent of Aquila's stock. Once HudBay spent another $10 million on exploration on the Back Forty project, they would gain control of the project, with their share increasing to 51 percent. HudBay fulfilled their obligation in just over a year, gaining their majority ownership in September 2010.

As diamond drilling continued at the Back Forty, the known ore resource continued to grow. Several types of mineralization were found in the ore body. The zinc-rich and copper-rich massive sulfide ore consists of aggregates of pyrite, sphalerite, chalcopyrite and galena. Where the massive sulfide ore veins are near the surface and exposed to the effects of atmospheric oxygen, some of the ore has oxidized into gossan, made up of the iron minerals hematite and goethite. Gold is found in varying grade in all of the ore types, as well as in some host rock where it is found nearly free from sulfides. The gossans have been found to be particularly rich in gold and silver.

By 2012 Aquila and HudBay had enough data on the value of the Back Forty to release a preliminary economic assessment of the project. From the assessment, drawn up under the strict Canadian National Instrument 43-101 standard, it was projected that an open pit mine could be opened on the shallow portions of the deposit that would have a seven-year operating life and would produce about a million metric tons of ore per year. A significant amount of ore exists beneath the proposed open pit extending to depths of more than 2,000 feet, creating the potential for underground development following mining of the open pit.

The projected mine would be an open pit about 1,900 feet long, 1,500 feet wide and 400 feet deep. Waste rock taken out during the development of the mine would be stored for re-use at the end of the mine's life-cycle, when the pit would be refilled with waste rock mixed with enough limestone to neutralize any acid that might be formed from any residual sulfides. Ore in the pit would be blasted and trucked to an on-site mill facility. Here most of the ore would be ground and the zinc and copper sulfides separated from the waste rock in separate flotation circuits. The copper concentrate would carry most of the gold and silver values from the sulfide ore, which would be recovered by the smelters, which would be the mine's customers.

The gold- and silver-rich oxide ore from the gossans and other gold zones would be treated in a separate hydrometallurgical process. The ground ore would be mixed with a weak cyanide solution to dissolve the precious metals, much as was done at the Ropes Gold Mine in the 1980s, with the end product being doré bullion to sell to precious metal refiners. While gold is only a minor constituent of the ore body, its high value would make a 48 percent contribution to the mine's revenue stream.

Less than three months after the preliminary economic assessment was released, HudBay announced that it was suspending its exploration activities at Back Forty to allocate its resources toward other key development projects. As of April 2013, the project was sitting idle while Aquila and HudBay look for other funding opportunities to turn the Back Forty prospect into a producing mine.

Ninety miles north of the Back Forty is the Eagle Mine, a nickel-copper mine due to go into production in 2014. The discovery of Eagle took more than 35 years from the first hint of nickel to the discovery of an economic ore body.

An aeromagnetic survey of the central Upper Peninsula in the mid-1970s showed magnetic anomalies in the Yellow Dog Plains in northern Marquette County. It was 11 years before these anomalies were investigated further. Once it was realized that the largest of these anomalies was centered on one of the few rock outcrops on the plains, geologists took a closer look and found it to consist of an igneous rock called peridotite.

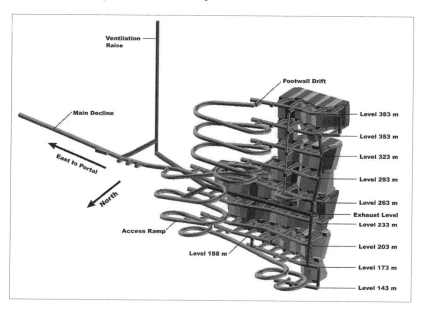

A diagram of the Eagle Mine

ALL IMAGES AUTHOR'S COLLECTION UNLESS INDICATED

During 1976, the Geological Survey Division of the Michigan Department of Natural Resources studied the Yellow Dog Plains area to see if the magnetic anomalies indicated a potentially mineral-rich complex of igneous rock. Using ground-based magnetic, gravity and very-low-frequency electromagnetic technologies, the geologists refined the location and estimated size of the mineralized bodies hidden beneath the plains. The same year, Michigan Technological University sank a 100-foot diamond-drill hole into the outcrop now known as Eagle Rock. Only low percentages of nickel and copper were detected. The Geological Survey's conclusion was that "the Yellow Dog Peridotite may be a favorable host rock for copper-nickel mineralization."

Eagle Decline Drilling
RIO TINTO

From 1991 to 1994, Kennecott Exploration prospected the Yellow Dog Plains and the surrounding area for zinc deposits. In the course of these explorations (which didn't uncover any significant amount of zinc ore), one of its geologists found boulders of peridotite carrying massive nickel and copper ore in the rip-rap on a bridge approach. The peridotite in the boulders was much richer than the exposed areas of the Yellow Dog Peridotite, convincing the geologist that there must be an economic deposit of massive sulfide somewhere in the vicinity.

In 1995 Kennecott Exploration shifted its focus to looking for magmatic nickel deposits. It drilled four diamond-drill holes into Eagle Rock, but only found disseminated sulfides too low-grade to mine. Its explorations in the area resumed in 2001, still targeting the peridotite body underlying the eastern outcrop. In the spring of 2002, Kennecott Exploration moved one of their diamond-drill rigs to the western outcrop, a mere bump of rock barely rising above the sand in the valley of the Salmon Trout River. The first diamond-drill hole into the peridotite below this unassuming outcrop passed through 270 feet of massive sulfide carrying 6.3 percent nickel and 4 percent copper – world-class ore.

Over the next few years, hundreds more diamond-drill holes were drilled, first by Kennecott Exploration, then by Kennecott Minerals Copper Division. The drilling and sampling delineated an intrusion of igneous rocks 1,575 feet long and 330 feet wide near the surface that tapers down to about 30 feet wide, 1,100 feet down. The formation carries a variety of sulfide minerals including pyrrhotite, an iron/nickel mineral; chalcopyrite, a copper ore; and pentlandite, a major ore of nickel. Other metals are found in tiny amounts in the ore, including cobalt, platinum, palladium and gold. The ore was apparently deposited in a wide spot in a magma tube that flowed molten rock toward the earth's surface during three successive eruption stages about 1.1 billion years ago.

Once the ore body was determined to be economically viable, the long process of obtaining all the necessary permits from Michigan Department of Natural Resources, Department of Environmental Quality and the U.S. Environmental Protection Agency began. Initial permit applications were filed early in 2006, and after the legal hurdles of lawsuits, court orders and injunctions were cleared, the final permit was issued in January 2010. Surface construction of the mine facility began that spring, but it wasn't until September 2011 that underground development began.

The mine was developed by driving a decline from the surface into the eastern peridotite intrusion, then one mile at a steady grade of 13 degrees down to the ore body. To help to preserve the natural state of Eagle Rock, which is a sacred site for the Keweenaw Bay Indian Community, the decline penetrates the intrusion well below surface, leaving the rock itself untouched. Once the decline reached the ore body, development levels were excavated, starting at the lowest draw level nearly 1,000 feet below surface. The mining method used at the Eagle Mine is known as long-hole stope mining. Production levels are driven in the country rock next to the ore body, then a series of parallel crosscuts are excavated under the ore to be mined. A matching set of crosscuts are driven on the mining level some 100 feet above the production level. Holes are drilled into the ore from the mining level, then are charged with explosives and detonated, breaking the ore to a manageable size. The broken ore is removed from the stopes by bucket loaders on the production levels and is loaded into 40-ton capacity underground haul trucks and transported up the decline to the covered ore storage building on surface.

Once a stope is mined out, it is packed with cemented rock fill to support the workings above. When a pair of stopes have been filled with concrete, the ore between them can be mined. The empty stopes between the concrete pillars are filled with waste rock. The process is repeated until all the ore has been mined from a level, then the mining moves to the next level up. Once the ore body is mined out and the mine closed, it will have been replaced by a solid body of concrete and waste rock.

Ore from the mine will be trucked to Humboldt, Michigan, where Kennecott Minerals, now doing business under the name of its parent company, Rio Tinto, has rehabilitated the Humboldt Mill. This facility was originally built in the 1950s to beneficiate low-grade iron ore from the Humboldt mine and was re-purposed in the 1980s to extract gold and silver from the Ropes Gold Mine.

At the Humboldt Mill, the ore is crushed down to ½-inch size in three stages of crushers, then ground in ball mills to the consistency of fine sand. The valuable minerals in the ore are then separated in flotation cells. Chemical reagents that adhere to the metallic minerals are added to the ore slurry. When air is bubbled through the slurry, the ore minerals cling to the bubbles and are floated off, while the mineral-poor portion of the ore sinks to the bottom and is disposed of as tailings. Subsequent stages of flotation separate the ore into a nickel concentrate and a copper concentrate, which are dried and shipped by rail to smelters for further refining.

The tailings left from the concentrating process are

deposited in the old water-filled Humboldt mine pit, as the Ropes tailings were. Water from the pit is recycled for process water in the plant; any water leaving the property is purified in a water treatment plant before being discharged into the Escanaba River.

In July 2013, Eagle was sold to Toronto-based Lundin Mining Corporation. Once production begins late in 2014, the Eagle Mine is expected to produce 300 million pounds of nickel and 250 million pounds of copper over the eight year life of the mine. It is also expected to produce more than 20,000 ounces of gold as well as 60,000 ounces of platinum, 40,000 ounces of palladium and 3300 tons of cobalt as byproducts recovered from the concentrates.

Although gold is only a minor constituent of the Eagle Mine's production, the opening of this mine is proof that a new mine can be developed and operated under today's strict environmental regulations. Modern explorers continue to search the Upper Peninsula for gold and silver, keeping alive the prospectors' dreams from a century and a half ago.

# Glossary

| | |
|---|---|
| ADIT | A horizontal or nearly horizontal entrance to a mine. |
| AMALGAM | An alloy of mercury with gold and/or silver. |
| ASSAY | Analysis of an ore to determine its composition and value. |
| BENEFICIATE | To treat raw materials, such as iron ore, to improve its physical or chemical properties in preparation for additional processing. |
| BULLION | Gold or silver in the form of bars or bricks. |
| CHALCOCITE | A gray or black sulfide of copper. |
| CHALCOPYRITE | A yellow sulfide of copper and iron. |
| CHIMNEY | A vertical rich streak in a vein. |
| CONCENTRATE | A product containing the valuable metal from which most of the waste material in the ore has been eliminated. |
| COUNTRY ROCK | The rock that has no mining value surrounding an ore body. |
| CROSSCUT | A horizontal underground opening driven across the strike of the vein. |
| DIAMOND DRILL | A drill having a hollow, cylindrical bit set with diamonds, used for obtaining cores of rock samples. |
| DIP | The vertical angle formed by a vein and the surface. |
| DORÉ | Unrefined gold bullion containing other metals. |

| | |
|---|---|
| DRIFT | A horizontal passage running through or parallel to a vein. To excavate such a passage. |
| FLOAT COPPER | Loose pieces of native copper that have been moved from their original location by glacial action. |
| FORTY | A 40-acre parcel of land, generally a quarter of a mile square. |
| FREE GOLD | Gold that can be extracted from the ore by amalgamation with mercury. |
| GALENA | A sulfide of lead, the most common lead ore. |
| GRIZZLY | A grating placed over the top of a chute for the purpose of stopping the larger pieces of ore from dropping through it. |
| HEADFRAME | The framework over a mine shaft that supports the hoisting apparatus. |
| LEVEL | The horizontal passages at a given depth in a mine. |
| LODE | A body of mineral deposited between clearly defined rock walls; a vein. |
| MANGANESE | A metallic element used in steel making. |
| MILL | A plant in which ore is treated to recover the valuable metals. Also a machine to grind the ore. |
| OPTION | The exclusive right to explore a parcel of land for minerals, or to purchase the land for a specified price. |
| PLACER | A deposit of sand or gravel containing minerals eroded out of their original occurrence. |
| PROSPECT | A mineral exploration site. To explore for minerals. |
| PYRITE | A yellow sulfide of iron, commonly known as "fool's gold." |
| QUARTZ | A hard, glassy rock in which gold is often found. |
| RIFFLES | Grooves or ridges in the bottom of a sluice designed to collect particles of gold. |

| | |
|---|---|
| SALT | To artificially enrich a mineral deposit. |
| SECTION | A mile-square parcel of land. |
| SHAFT | A vertical or steeply angled entrance to a mine. |
| SILVER LEAD | Galena (a lead ore) containing silver. |
| SKIP | A self-dumping hopper on wheels, used to hoist ore up a shaft. |
| SLUICE | A trough in which gold is separated from sand and gravel by the action of water. To wash in a sluice. |
| SMELTER | A furnace for extracting metal from ore by the application of heat. |
| STAMP MILL | A machine in which ore is crushed to powder by heavy stamps or pestles. |
| STOPE | An opening from which ore is removed. |
| STRIKE | The direction of the surface exposure of a vein. |
| SUGAR QUARTZ | A white, granular form of quartz. |
| SULFIDE | A compound of sulfur and a metal, often an ore of the metal. |
| TAILINGS | The waste rock left after the economically valuable minerals have been extracted. |
| TAILRACE | The channel for conducting tailings away in water. |
| TALCOSE | Containing the soft, soapy mineral talc. |
| TELLURIUM | A silvery white metal that sometimes occurs in combination with gold. |
| TRAM | A small wheeled car running on rails used to transport ore. |
| TROY | A system of weights used to weigh precious metals. (A troy ounce is equal to 1.097 ordinary or avoirdupois ounces.) |
| VANNER | An inclined vibrating rubber belt used to concentrate heavy mineral ores. |
| VEIN | A long, narrow body of mineral; a lode. |

# Bibliography

Allen, R.C., "Gold in Michigan." *Mineral Resources of Michigan with Statistical Tables of Production and Value of Mineral Products for 1910 and Prior Years Michigan Geological and Biological Survey, Publication 8, Series 6*, p. 355-366, Lansing, Michigan: State Printers, 1912

Banfield, J.A., New & Correct Sectional Map of the Iron, Silver & Gold Region, Lake Superior, Michigan, 1864.

Barnett, LeRoy, *Mining in Michigan: A Catalog of Company Publications 1845-1980*. Marquette, Michigan: Northern Michigan University Press, 1983.

Barton, P.B. Jr., and Simon, F.O., "Gold Content of Michigan Native Copper." Geological Survey Research 1972, Chapter A U.S. Geological Survey, Professional Paper 800 A, p. A 4.

Baxter, D.A., Bornhorst, T.J. and VanAlstine, J.L., "Geology, Structure, and Associated Precious Metal Mineralization of Archean Rocks in the Vicinity of Clark Creek, Marquette County, Michigan." Open File Report OFR 87 8, Geological Survey Division, Michigan Department of Natural Resources, 1987.

*Beard's Directory and History of Marquette County*. Detroit, Michigan: Hadger & Bryce, Steam Book Printers, 1873.

Benedict, C. Harry, *Lake Superior Milling Practice*. Houghton, Michigan: The Michigan College of Mining and Technology, 1955.

*Biographical Record.* Chicago: Biographical Publishing Company, 1903.

Boben, C.L., Bornhorst, T.J. and VanAlstine, J.L., "Detailed Geological Study of Three Precious Metal Prospects in Marquette County and One in Gogebic County, Michigan." Open File Report OFR 86 1, Geological Survey Division, Michigan Department of Natural Resources, 1986.

Bodwell, Willard Arthur, "Geological Compilation and Nonferrous Metal Potential of the Precambrian Section of Northern Michigan MS Thesis." Michigan Technological University, 1972.

Bornhorst, Theodore J., Shepeck, Anthony W. and Rossell, Dean M., "The Ropes Gold Mine, Marquette County, Michigan, U.S.A. An Archean Hosted Lode Gold Deposit," in MacDonald, A.J., ed. *Proceedings of Gold '86, an International Symposium on the Geology of Gold* p. 213-227. Toronto: 1986.

Bradish, Alvah, *Memoir of Douglass Houghton.* Detroit: Raynor & Taylor, Printers, 1889.

Brewer, Carl, *Silver-Lead Explorations in Marquette County.* Unpublished manuscript, n.d.

*A Brief History of the Pittsburgh and Lake Superior Iron Company.* Pittsburgh: The Pittsburgh and Lake Superior Iron Company, 1883.

Broderick, T.M., "Geology of the Ropes Gold Mine." *Bulletin of the Michigan College of Mining and Technology*, Houghton, Michigan: June 1945.

Brooks, T.B., *Geological Survey of Michigan, Upper Peninsula 1869-1873, Volume I & II.* New York: Julius Bien, 1873.

Brozdowski, Robert A., Gleason. Richard J. and Scott, Glenn W., "The Ropes Mine: A Pyritic Gold Deposit in Volcaniclastic Rock, Ishpeming, Michigan, U.S.A." in MacDonald, A.J., ed. Proceedings of Gold '86, an International Symposium on the Geology of Gold, p.

228-242, Toronto: 1986

Callahan Mining Corporation, Annual Reports, 1980-1987.

Cannon, W.F., "Mineral Resource Assessment of the Iron River 1° X 2° Quadrangle." *Michigan and Wisconsin Geological Survey Circular 887*, 1983.

Cannon, W.F., King, Elizabeth R., Hill, James J. and Morey, Peter C., *Mineral Resources of the Sturgeon River Wilderness Study Area, Houghton and Baraga Counties, Michigan*. Washington: U.S. Government Printing Office.

Cannon, William F. and Klasner, John S., "Bedrock Geologic Map of the Southern Part of the Diorite and Champion 7 ½ Minute Quadrangles, Marquette County, Michigan." U.S. Geological Survey Miscellaneous Investigations Series Map I 1058, 1977.

Carlson, Shawn M., Robinson, George W., Elder, Mark J, Jaszczak, John A., and Bornhorst, Theodore J., "Greenockite and Associated Uranium-Vanadium Minerals from the Huron River Uranium Prospect Baraga County, Michigan." *Rocks and Minerals, Volume 82*, July/August 2007.

Carter, James L. and Rankin, Ernest H., eds. *North to Lake Superior: the Journal of Charles W. Penny 1840*. Marquette, Michigan: The John M. Longyear Research Library, 1970

*The Chicago Times*. Chicago, Illinois, March 5, 1864.

Clark, L.D., Cannon, William F., and Klasner, John S., "Bedrock Geologic Map of the Negaunee SW Quadrangle, Marquette County, Michigan." U.S. Geological Survey Geologic Quadrangle Map GQ 1226, 1975.

Clarke, Don H., *The Gold Mines of Michigan*, 1976.

Courtis, W.M., "Gold in Michigan." Michigan Geological Service Annual Report, 1906, p. 581-584.

Cox, Bruce K., *Mines of the Pewabic Country of Michigan and Wisconsin Volume 1 Gold and Silver*. Marquette, Michigan: Agogeebic Press, 2002.

Denning, R.M., "Geology of the Ishpeming Gold Range." Michigan Geological Survey Open File Report, 1948.

*Detroit Free Press*, Detroit, Michigan, June 14, July 3, 1889.

*Detroit News*, Detroit, Michigan, August 18, 1935.

Ding, Xin, Ripley, Edward M., Li, C., "Geochemical and stable isotope studies of hydrothermal alteration associated with the Eagle deposit, northern Michigan." Proceedings of the Institute on Lake Superior Geology, v. 54, Part 1 – Program and Abstracts, 14-15.

Ding, Xin, Edward M. Ripley and Chusi Li (2012) "PGE geochemistry of the Eagle Ni–Cu–(PGE) deposit, Upper Michigan: constraints on ore genesis in a dynamic magma conduit." *Mineralium Deposita* 47:89-104.

Disturnell, J., *The Great Lakes, Inland Seas of America*. New York: Charles Scribner, 1865.

Dorr, J.A., and Eschman, D.F., *Geology of Michigan*. Ann Arbor, Michigan: The University Press, 1970.

Edwards, James P., *Mines and Mineral Statistics*, p. 113 115, Lansing, Michigan: Robert Smith and Company, 1892.

*The Engineering and Mining Journal*, various issues. New York: The Scientific Publishing Company.

*Evening News*, Detroit, Michigan, August 28, 1888.

"First Annual Report of the Trustees of the New-York and Lake Superior Mining Company." Albany, New York: The Evening Atlas, February 24, 1846.

Foster, J.W. and Whitney, J.D., "Report on the Geology and Topography of a Portion of the Lake Superior

Land District in the State of Michigan." Washington: House of Representatives, 1850.

Fuller, George N., ed., *Geological Reports of Douglass Houghton*. Lansing, Michigan: The Michigan Historical Commission, 1928.

Gair, Jacob E. and Thaden, Robert E., "Geology of the Marquette and Sands Quadrangles, Marquette, Michigan." *U.S. Geological Survey Professional Paper 397*, 1968.

"The Gold Fields." *The Engineering and Mining Journal*, August 18, 1888.

"Gold in Michigan Open File Report MGSD OFR GOLD 80 1." Geological Survey Division, Michigan Department of Natural Resources, 1980.

"Gold Mines of Michigan." *The New York Times*, August 27, 1888.

Gray, A.B., "Report of A.B. Gray Relative to the Mineral Lands on Lake Superior." Dated March 10, 1846. Transmitted to the House of Representatives, June 16, 1846. 29th Congress, 1st session, 1845-46. Executive Document, Vol. VII, No. 211.

Hawke, Richard, *A Guide to Rocks and Minerals of Michigan*. Midland, Michigan: Dick Hawke Science Service, 1981.

Heinrich, E. William, *The Mineralogy of Michigan Geological Bulletin 6*. Lansing: State Printers, 1976.

Heinrich, E.W. and Robinson, George W., *Mineralogy of Michigan* (Second Edition). Houghton, Michigan: A.E. Seaman Mineral Museum, 2004.

Henry, Alexander, *Travels & Adventures in Canada and the Indian Territories*. Boston: Little, Brown & Company, 1901.

"History of the Discovery and Workings of the Silver Mines in the Iron River District Ontonagon County." Ontonagon, Michigan: *Ontonagon Miner*, 1875.

*History of the Upper Peninsula of Michigan*. Chicago: The Western Historical Company, 1883.

Holland, A.H., *1891 Handbook and Guide to Ishpeming, Michigan, Together with a Mining Directory of Marquette County.*

Holland, A.H., *A Gazetteer of Marquette County.* Marquette, Michigan: Mining Journal Print, 1889.

Holland, A.H., *The Marquette City Directory*. Marquette, Michigan: Mining Journal Print, 1891.

Holland, A.H. and Dwight, E.H., *1886-7 Handbook and Guide to Ishpeming, L.S., Michigan*. Marquette, Michigan: Mining Journal Book and Job Print, 1886.

Houghton, Jacob, *The Mineral Region of Lake Superior*. Buffalo, New York: Oliver G. Steele, 1846.

Houghton, J. Jr. and Bristol, T.W., *Reports of Wm. A. Burt and Bela Hubbard, Esqs. on the Geography, Topography and Geology of the U.S. Surveys of the Mineral Region of the South Shore of Lake Superior for 1845*. Detroit: Charles Willcox, 1846.

Hubbard, Lucius L., *Geological Survey of Michigan, Upper Peninsula 1881-1884, Lower Peninsula 1885-1893, Volume V.* Lansing, Michigan: Robert Smith & Co., 1895.

*The Iron Agitator*, Ishpeming, Michigan, various issues 1881-86.

*The Iron Herald*, Negaunee, Michigan, various issues 1884-94.

*The Iron Home*, Ishpeming, Michigan, various issues 1874.

*The Iron Ore*, Ishpeming, Michigan, various issues 1886-1953.

"The Ishpeming Gold Range," *The Engineering and Mining Journal*, July 5, 1890.

Jamison, James K., *The Mining Ventures of This Ontonagon*. Ontonagon, Michigan: by the author, ca. 1950.

Jamison, James K., *This Ontonagon Country: The Story of an American Frontier*. Ontonagon, Michigan: The Ontonagon Herald Company, 1939.

Jamison, Knox, *A History of Silver City, Ontonagon County Michigan*. Silver City, Michigan: by the author, 1963.

Johanson, Bruce H., *This Land, the Ontonagon: A Short History of Ontonagon County, Michigan*.

Johnson, R.C., Bornhorst, T.J. and VanAlstine, J.L., "Geology and Precious Metal Mineralization of the Silver Creek to Rocking Chair Lakes Area, Marquette County, Michigan Open File Report OFR 86 2." Geological Survey Division, Michigan Department of Natural Resources, 1986.

Jopling, James E., "Personal Reminiscences of a Mining Engineer." *Michigan History Magazine*, April 1927.

Kelly, W.A., "Economic Geology of the Dead River Area." Unpublished manuscript (report to the Norgan Gold Mining Company), 1936.

Kelly, W. A., "Geology of the Dead River Area, Marquette County, Michigan." Unpublished manuscript (report to the Norgan Gold Mining Company), 1936.

Klasner, John S., Snider, David W., Cannon, W.F. and Slack, John F., "The Yellow Dog Peridotite and a Possible Buried Igneous Complex of Lower Keweenawan Age in the Northern Peninsula of Michigan." Lansing, Michigan: Department of Natural Resources, Geological Survey Division, 1979.

Knight, James B., *Mines and Mineral Statistics*, p. 159 160. Lansing, Michigan: Robert Smith and Company, 1894.

Koschman, A.H. and Bergendahl, M.H., *Principal Gold Producing Districts of the U.S. Geological Survey Professional Paper 610.* Washington: Government Printing Office, 1962.

Kronquist, E.A., "Report on Exploration of Section 35, T49N R27W, Michigan." Unpublished manuscript (report to the Norgan Gold Mining Company), 1936.

LaFayette, Kenneth, *Marquette Gold and Silver Finds.* Unpublished manuscript, Marquette Historical Society, 1975.

*Lake Superior Journal*, Sault Ste. Marie, Michigan, August 7, 21, 1850; April 1, September 15, 1852; April 5, 1857; January 2, 1964.

*Lake Superior News and Mining Journal*, Copper Harbor, Michigan, August 8, 22 and 29, 1846.

*Lake Superior Silver-Lead Company.* New York: Howe & Ferry, Stationers, 1864.

Lamey, Carl A., *Lead-Silver Veins of Michigan.* Unpublished manuscript, Geology Department, Michigan College of Mining and Technology, 1935.

Lamey, Carl A., *Michigan Gold.* Unpublished manuscript, Geology Department, Michigan College of Mining and Technology, 1935.

Lamey, Carl A., "Notes Regarding Geological Conditions Observed During Summer of 1936, Norgan Gold Mining Company, Section 35, T47N R27W." Unpublished manuscript (report to the Norgan Gold Mining Company), 1936.

Lane, Alfred C., "Report of the State Board of Geological Survey of Michigan for the Year 1904," p. 156-158. Lansing, Michigan: 1905.

Lane, Alfred C., "Report of the State Board of Geological Survey of Michigan for the Year 1906," p. 581-584. Lansing, Michigan: 1907.

"Large Nuggets of Gold," *The New York Times*, July 21, 1888, p. 1.

Lawton, Charles D., *Mineral Resources*, p. 157-161. Lansing, Michigan: Thorp and Godfrey, 1886.

Lawton, Charles D., *Mines and Mineral Statistics*, p. 269-273. Lansing, Michigan: Thorp and Godfrey, 1887.

Lawton, Charles D., *Mines and Mineral Statistics*, p. 137-143. Lansing, Michigan: Thorp and Godfrey, 1888.

Lawton, Charles D., *Mines and Mineral Statistics*, p. 91-100. Lansing, Michigan: Darius D. Thorp, 1889.

Lawton, Charles D., *Mines and Mineral Statistics*, p. 65-69. Lansing, Michigan: Robert Smith and Company, 1890.

Lawton, Charles D., *Mines and Mineral Statistics*, p. 83. Lansing, Michigan: Robert Smith and Company, 1891.

Learned, William Law, *The Learned Family: (Learned, Larned, Learnard, Larnard and Lerned) Being Descendants of William Learned, who was of Charlestown, Massachusetts, in 1632*. Albany, New York: J. Munsell's Sons, 1882.

Lehto, Steve, *Michigan's Columbus: The Life of Douglass Houghton*. Royal Oak, Michigan: Momentum Books, 2009.

Lundin Mining Corporation, "NI 43-101 Technical Report on the Eagle Mine, Upper Peninsula of Michigan, USA," Wardell Armstrong, Office S15, Tremough Innovation Centre, Penryn, Cornwall, TR10 9TA. July 2013.

Margeson, G.B., Norby, J.W., Brozdowski, R.A., Carter, A.S. and Bouley, B.A., "Comparison of Two Parts of the Dead River Ishpeming Greenstone Belt: Evidence for Correlation of Volcanic Stratigraphy" (abs.). Institute on Lake Superior Geology Proceedings and Abstracts, V. 34, Part I, p. 70, 1988.

McGhee, George W., "Vanadium Discovery in Baraga County, Mich." *Mining and Engineering World*, December 12, 1914.

"The Michigan Gold Fields," *The New York Times* September 2, 1888, p. 9.

"The Michigan Gold Mines," *The New York Times*, July 30, 1888, p. 5.

*Michigan State Gazetteer and Business Directory for 1863-4*. Detroit, Michigan: Charles F. Clark, 1863.

*Michigan State Gazetteer and Business Directory, Volume V*. Detroit, Michigan: R.L. Polk & Co., 1881.

*Michigan State Gazetteer and Business Directory, Volume VII*. Detroit, Michigan: R.L. Polk & Co., 1885.

*Michigan State Gazetteer and Business Directory, Volume VIII*. Detroit, Michigan: R.L. Polk & Co., 1887.

*Michigan State Gazetteer and Business Directory, Volume X*. Detroit, Michigan: R.L. Polk & Co., 1891.

*Mineral Processing Flowsheets*. Denver: Denver Equipment Company, 1962.

*Mining Journal*, Marquette, Michigan, various issues 1881-1990.

Nankervis, James L., *Mines and Mineral Statistics*, p. 218. Houghton, Michigan: Gazette Print, 1908.

*The New York Times*, July 16, 1887, and July 26, August 28, September 7 and 10, 1888.

Newett, George A., *Mines and Mineral Statistics*, p. 164-170. Lansing, Michigan: Robert Smith and Company, 1896.

Newett, George A., *Mines and Mineral Statistics*, p. 169-172. Ishpeming, Michigan: Iron Ore Printing House, 1897.

Newett, George A., *Mines and Mineral Statistics*, p. 200, 215-220. Ishpeming, Michigan: Iron Ore Printing House, 1898.

Newett, George A., *Mines and Mineral Statistics*, p. 289, 299-302. Ishpeming, Michigan: Iron Ore Printing House, 1899.

Newett, George A., "A Michigan Gold Mine." *Michigan History* Magazine, Vol. 11, No.1, p. 73-91, 1927.

Norby, J.W., 1988, "History of Precious Metal Exploration/Development in the Dead River Ishpeming Greenstone Belt" (abs.) Institute on Lake Superior Geology Proceedings and Abstracts, V. 34, Part I, p. 82-84.

Owens, E.O. and Bornhorst, T.J., "Geology and Precious Metal Mineralization of the Fire Centre and Holyoke Mines Area, Marquette County, Michigan Open File Report OFR 85 2." Geological Survey Division, Michigan Department of Natural Resources, 1985.

Pardee, Franklin G., "Michigan's Mythical Gold Mines." *Michigan Conservation*, Vol. 14, No. 11, p. 5, 9 10, November, 1945.

Parker, Richard A., "The New Michigan Gold Finds." *The Engineering and Mining Journal*, p. 238-239, September 22, 1888.

*Portage Lake Mining Gazette*, Houghton, Michigan, various issues.

Powers, H.M., "Romance and Adventure on the Ontonagon," *Michigan History*, Vol. 5, 1921.

"Prospectus for the Formation of the Fortuna Mining Company of Michigan, Section 16 T49N-R28W." Philadelphia, 1864.

"Prospectus for the Formation of a Silver-Lead Mining Company on Section 14, Town 49 North, Range 28 West, Lake Superior, Michigan." Philadelphia: J.B. Chandler, Printer, 1863.

"Prospectus of the Pioneer Silver and Lead Company." Detroit: Detroit Free Press Steam Book and Job Printing Establishment, 1864.

Puffett, Willard P., "Geology of the Negaunee Quadrangle, Marquette County, Michigan." *U.S. Geological Survey Professional Paper 788*, 1974.

Puffett, Willard P., "Occurrences of Base Metals South of Dead River, Negaunee Quadrangle, Marquette County." Michigan 12th Annual Institute of Lake Superior Geology, p. 18, 1966.

Quirke, T.T., "Preliminary Geological Report, T49N R27W Section 35, Marquette County, Michigan." Unpublished manuscript (report to the Norgan Gold Mining Company), 1936.

Rainbow Exploration Company, "Geology of the Eagle Mills Morgan Area." Unpublished geological map, ca. 1988.

"Rich Gold Deposits Found in Michigan," *The New York Times*, December 8, 1935.

Robers, W. Chandler, "Gold." *Encyclopedia Britannica*. Chicago: The Werner Company, 1894.

Robinson, Orrin W., *Early Days of the Lake Superior Copper Country*. Houghton, Michigan: D.L. Robinson, 1938.

Rominger, C., *Geological Survey of Michigan, Lower Peninsula 1873-1876, Volume III*. New York: Julius Bien, 1876.

Rominger, C., *Geological Survey of Michigan, Upper Peninsula 1878-1880, Volume IV*. New York: Julius Bien, 1881.

Ropes Gold and Silver Company, Annual Reports, 1885-1894.

Rossell, Dean and Coombes, Steven, "The Geology of the Eagle Nickel-Copper Deposit Michigan, USA." prepared for Kennecott Minerals Co., 2005.

Rossell, Dean and Kalliokoski, J., "Ropes Gold Mine and its Geological Setting," in Proceedings 29th Annual Institute on Lake Superior Geology. Houghton,

Michigan: Michigan Technological University, 1983.

Russell, James, *Mines and Mineral Statistics*, p. 127. Marquette, Michigan: Mining Journal Company, Ltd., 1901.

Rydholm, C. Fred, *Superior Heartland: A Backwoods History*. Marquette, Michigan: by the author, 1989.

St. John, John R., *A True Description of the Lake Superior Country, its Rivers, Coasts, Bays, Harbours, Islands, and Commerce*. New York: William A. Graham, 1846.

Schmeling, E.S., "Report on T49N R28W, Sections 1, 2, 3, 10, 11, 12, 13, 14 & 15, T49N R27W, Sections 6, 7 & 18." Unpublished manuscript (report to the Norgan Gold Mining Company), 1936.

Seagall, R. Tom, and Seagall, Glenna, "14 Years of 24 Karat Mining." *Michigan Natural Resources*, Vol. 44, No. 6, p. 7-9, November/December 1975.

Seeland, David A., "A Geochemical Reconnaissance for Gold in the Sedimentary Rocks of the Great Lakes Region, Minnesota to New York." *U.S. Geological Survey Bulletin 1305*, 1973.

Skillings, David N., "Callahan Begins Development of its Ropes Gold Mine for 1985 Operation." *Skillings Mining Review*, September 17, 1983.

Snelgrove, A.K., Seaman, W.A. and Ayres, V.L., "Strategic Minerals Investigations in Marquette and Baraga Counties, 1943." Michigan Geological Survey Progress Report No. 10, 1944.

Snider, David W., "Investigation of a Reported Copper Silver Showing in Section 30, T50N, R26W, Marquette County, Michigan Open File Report OFR 77 2." Geological Survey Division, Michigan Department of Natural Resources, 1977.

Sprague, Marshall, *Money Mountain; The Story of Cripple Creek Gold*. Boston: Little, Brown and Company, 1953.

Stonehouse, Frederick, *Marquette Shipwrecks*. AuTrain, Michigan: Avery Color Studios, 1977.

Swineford, Alfred P., "Appendix to Swineford's History of the Lake Superior Iron District, Being a Review of its Mines and Furnaces for 1873." Marquette, Michigan: *The Mining Journal*, 1873.

Swineford, Alfred P., "Annual Report of the Commissioner of Mineral Statistics of the State of Michigan for 1883," p. 98-100, 112-117. Marquette, Michigan: Marquette Mining Journal Publishing House, 1884.

Swineford, Alfred P., "Annual Report of the Commissioner of Mineral Statistics of the State of Michigan for 1884," p. 9-11. Marquette, Michigan: Marquette Mining Journal Publishing House, 1885.

Swineford, Alfred P., "Annual Review of the Iron, Copper and Other Industries of the Upper Peninsula of Michigan for the Year Ending Dec, 31, 1882." Marquette, Michigan: *Marquette Mining Journal*, 1883.

Swineford, Alfred P., "History and Review of the Copper, Iron, Silver, Slate and Other Material Interests of the South Shore of Lake Superior." Marquette, Michigan: *The Mining Journal*, 1876.

Thirtell, Joel, "Gold Fever," *Detroit Free Press*, Detroit, Michigan, May 15, 1991.

Thwaites, Reuben Gold, LL. D., *Collections of the State Historical Society of Wisconsin, Volume XVII The French Regime in Wisconsin – II, 1727-1748*. Madison: State Historical Society of Wisconsin, 1906.

Tyler, Stan, "Sections 8, 9, Northern Half of 16 & 17, etc. T49N R27W." Unpublished manuscript (report to the Norgan Gold Mining Company), 1936.

Wadsworth, M.E., "A Sketch of the Geology of the Iron, Gold, and Copper Districts." Report of the

State Board of Geological Survey for the Years 1891 and 1892, Michigan Geological Survey, p. 152-155, Lansing, Michigan, 1893.

Wakefield, Larry, *Ghost Towns of Michigan, Volume 2*. Thunder Bay, Ontario: Thunder Bay Press, 1995.

*Wakefield Bulletin*, Wakefield, Michigan, various issues 1887-1888.

Ware, Andrew, Cherry, Jon and Ding, Xin, "Geology of the Eagle Project." Proceedings of the Institute on Lake Superior Geology, v. 54, Part 2 – Field Trip Guidebook, 2008.

Wayment, Ross, Ropes unit manager, Callahan Mining Corporation, interview, February 7, 1985.

Willson, James Grant and Fiske, John, eds., *Appleton's Cyclopaedia of American Biography*. New York: D. Appleton and Company, 1889.

Wright, Charles E., "First Annual Report of the Commissioner of Mineral Statistics of the State of Michigan for 1877-8 and Previous Years." Marquette: Mining Journal Steam Printing House, 1879.

Zinn, Justin, "Field Report Covering the Field Mapping in 1936 for the Norgan Gold Mining Company." Unpublished manuscript (report to the Norgan Gold Mining Company), 1936.

# Index

**A**
Ada Creek, 192
Adams, John Q., 123, 128, 134, 136
Agate Harbor, 7
Agawam, Mass., 34
Agawam Silver Mining Co., 34
Agogebic Exploring, Mining and Manufacturing Co., 177, 178
Alaska, 29, 82
Albany, N.Y. 5
Alburtis, Edward, 14
Alcona Mining Co., 17, 24
Alger Gold Mining Co., 128
Alger, Russell A., 68, 128, 129
Alio, John, 79
Allegan County, 192
Allen, R.C., 172
Allen, Tom, 32
Allen, W.A., 135
Allerton, David, 14
Alpena, Mich., 198, 199
Amazon Mining Co., 120
American Gold and Silver Lead Mining Co., 19
Amerman, A.S., 50
Ames, James, 135
Ames, Joseph A., 123
Andersen, Nels, 194
Anderson, A.A., 129
Anderson, Andrew, 156
Anglo American Land and Mineral Co., 135
Anthony, E.C., 43
Antigo, Wis., 184
Antrim County, 192
Aquila Resources Inc., 108, 117, 141, 207
Arcadian Copper Mine Tours, 85
Archambeau, Alfred, 162
Argentine Silver Mining Co., 44
Arthur, George H., 132
Ashland Lumber Co., 48
Ashland, Wis., 36, 37, 48, 56
Atkinson, Henry, 170
Atlantic Dynamite Co., 80

Au Min Co. Inc., 195
Aurora, Ill., 66, 159
Aurora Smelting and Refining Co., 66, 159
Avery, C.A., 123

**B**
Back Forty, xx, 206, 207, 208, 209
Bacon, Roswell B., 124
Bagdad Pond, 147
Ball, Dan H., 155
Bancroft Iron Co., 21
Bangs, Nick, 184
Baraga County, 21, 22, 35, 149, 166, 167, 168, 169, 170, 173
Barnes, George, 161
Barnum, P.T., 133
Bateman, Will A., 104
Baxter, Alexander, 2
Bayfield, Wis., 56
Beanston, Peter C., 42, 54
Bear Creek, 159
Bear Creek Mining Co., 159, 203
Beardsley, Washington J., 43
Beaser, Daniel, 38
Beaver Board Co., 148
Beaver Granulith Co., 148, 149
Beechwood, Mich., 197
Begole, Frederick H., 134
Belanger, John B., 171
Bending, Ferdinand, 156
Bennett, E.P., 113, 123
Berringer, George, 127, 176
Bessemer, Mich., 179, 184, 190
Beth Cliffs Joint Venture, 202
Bethlehem Steel, 202
Big Bay, 14, 194
Big Garlic River, 164
Billings, James H., 133
Birmingham, Mich., 192
Bjork, Albert, 82, 202
Black Hawk War, 25
Blake, Edward, 128
Blake jaw crusher, 47, 63, 68, 100
Blake, Richard, 122, 156

*Bloodstoppers and Bearwalkers*, 196
B & M Gold Co., 134, 169
Boston, 109, 110, 123, 183
Bostwick, Henry, 2
Boulder City, Colo. 136
Boulsom, Abram, 130, 131
Boyne River, 192
Braastad, Frederick, 80, 111, 123
Bradish, Alvah, 3
Breitung, Edward, 113, 115, 118, 148
Briggs, Charles C., 43
Brix, E., 181
Broad, Josiah, 160
Broderick, T.M., 194
Bronson, Randall P., 134, 169
Brooks, William, 136
Brown, C.W., 58
Brown, W.H.A., 124
Bruce, Mich., 148
Brulé, Étienne, 2
Buchanan, John, 152
Buckley, H.J., 16
Buell, Andrew W., 16
Buell, Jonathan L., 174
Buffalo, N.Y., 148
Bughole Mine, 172
Bulldog Lake, 20
Burke, John, 76
Burr Oak, Mich., 192
Burt, Alvin C., 24
Burt, Hiram A., 24, 43
Burt, John, 12, 43
Burt, Samuel S., 23
Burt, William A., xiv, 4, 192
Burtis Sawmill, 157
Bush, I.D., 175
Butler, John P.M., 20
Butler, William, 58
Button, Charles F., 162
Byrne, F. Stuart, 161
Byrne, Samuel E., 152

232

## C

Callahan Mining Corp., ix, xix, 85, 94, 108, 112, 117, 126, 140, 203
Calumet, xix, 83, 84, 100, 104, 134
Calumet and Hecla Consolidated Copper Co., 83, 84, 100
Calumet Gold and Silver Mining Co., 134
Campau, Edward, 141
Campau, Frederick, 141
Campbell, John D., 140
Carey, M., 138
Cariboo Mining Co., 49
Caribou Island, 3
Carpenter, William T., 61
Carp River, 67, 76, 126, 136, 146, 162, 192
Carp River Furnace, 162
Carson, James, 14
Casa Grande, Ariz., 87
Cascade Iron Co., 11
Case, John R., 50
Case, Julian M., 115, 119, 158, 177
Centennial Mining Co., 49
Central Mining Co., 11
chalcocite, 9, 59, 158, 161
chalcopyrite, xiv, 7, 8, 19, 26, 65, 129, 132, 133, 136, 147, 151, 158, 161, 172, 174, 207, 211
Challenge disc feeders, 73
Champion, Mich., 50, 197, 203
Charlevoix County, 192
Cherokee Mining Co., 18, 34
Chevron Oil, 202
Chicago, 46, 48, 51, 53, 56, 58, 59, 114, 124, 126, 127, 128, 134, 139, 140, 150, 153, 154, 155, 158, 159, 160, 171, 172, 180, 181, 198, 200
Chicago Inter Ocean, 200
Chicago Northwestern Railroad, 150
Childs, Brainard F., 39
Ching, Christopher, 171
Chippewa Mining Co., 17, 27, 34, 167
Chocolay. Mich., see Harvey
Chocolay River, 5, 149
Christmas, Oliver, 158
Church, Sydney E., 25, 146
Cincinnati Gold and Silver Mining Co., 21, 28
Citizen's Fire Insurance Co., 14
Civilian Conservation Corps, 194
Civil War, 48, 128, 178, 184
Clark, Andrew G., 25, 33
Clark Creek, 202
Clark, Jeremiah, 9
Cleaves, W.S., 171, 172
Cleveland, xviii, 19, 50, 51, 52, 53, 54, 55, 56, 57, 80, 97, 100, 135, 148, 154, 165, 170, 202, 203

Cleveland Cliffs Iron Co., 135, 148, 165, 170, 202, 203
Cleveland Silver Mining Co., xviii, 51, 52, 53, 55, 56, 57
Clinton County, 195
Clintonville Mining and Exploring Co., 180, 185
Clintonville, Wis., 180
Clowry, Mich., 165
Cohodas, Samuel M., 107
Cold Springs, New York, 6
Collins, Charles G., 49, 51
Collins, Charles G., 41
Collins, L.H., 122
Comstock Gold and Silver Co., 186
Comstock, Sylvester, 13
Condon, William, 10
Conrad, C.F., 122
Considine, John Jr., 163
Conway, Martin, 171
Coon, Ephraim, 113, 123, 138, 165
Coon Gold Mining Co., 138
Cooper, Andrew J., 124
Cooper, James R., 100, 101
Copper Harbor, 5, 7, 19
Copper Range Mining Co., 90
Copps, Edward, 104, 176, 183
Corliss steam engine, 68
Cornish, 7, 32, 40, 42, 47, 57, 63, 73, 74, 157, 189
Cornwall, 27, 29, 63, 78
Corrigan, McKinney and Co., 80, 81, 82
Corser, Austin, xvii, 36, 37, 39, 41, 43, 44, 58
Corser, Sarah, 38, 43
Cosette Mining Co., 35, 166
Coulter, Joseph, 166
Courtis, William M., 116, 191, 192
Covington, Mich., 173
Cox, Jonathan, 20
Crary, Leonard P., 98
Crawford centrifugal ball mill, 73, 75, 144, 145
Cripple Creek, Colo., 120
Crocker, M.H., 43
Crosby, F.W., 47
Crow, Richard, 136, 153
Crystal Lake. see Bulldog Lake
Crystal Lake Silver Lead Mining and Smelting Co., 20, 27
Cummings, Frank, 67
Cummings, George P., 17, 21, 22, 23, 25, 61, 121, 146, 149
Cummings, Hannah, 61, 121
Cunningham, Walter, 5
Curry shaft, xviii, xix, xx, 62, 63, 68, 86, 91
Curry, Solomon S., 61, 62, 143

## D

Gust Dalman, 78
Daniel, Martin, 160

Davies, John, 125
Davis, Henry, 169
Davis, Theodore M., 119
Davis, William B., 10
Davis, William Morris, 17
Day, Samuel M., 17
Dead River, xi, xix, 7, 25, 34, 75, 112, 137, 139, 140, 142, 143, 145, 156, 165, 170, 193, 202, 203
Dead River Gold Mining Co., 140
Dead River Gold Range, xi, 137, 193
Dead River Miners' Pool, 142
Deer Lake, 61, 76, 81, 109, 124, 125, 126, 127
Deer Lake Furnace, 61, 76, 127
Deer Lake Iron Co., 81, 124
Denis, Louis, Sieur de la Ronde, 2, 36
Denniston, Garret V., 6
DePere, Wis., 116
Detective Mining Co., 181, 185
Detroit, 3, 4, 9, 14, 18, 19, 20, 22, 23, 28, 34, 41, 68, 79, 105, 116, 122, 141, 152, 155, 163, 166, 169, 175, 192, 198, 200
Detroit and Marquette Silver Lead Co., 34
Detroit City Silver Mining Co., 34
Detroit Free Press, 23
Detroit Gold and Silver Mining Co., 141
Dewey, W.C., 111
Dexter iron mine, 63
Dickinson County, xi, 166, 173
Doelle, John, 145
Dolf, Edward, 44
Donkersley, Cornelius, 24
Dooling, Thomas, 171
Dorrance, L.K., 24
Dorson, Richard M., 196
Doty, Duane, 23
Douglass, Columbus, 15
Douglass, Edward F., 15
Douglas, Silas H., 15
Duluth, ii, 5, 56, 147, 173, 244
Duluth, South Shore and Atlantic Railroad, 147, 148, 149, 173
Duncan, William C., 34
Dunkee, H.R., 124
Dunlap Mining Co., 182
Dwyer, James, 114, 170
Dwyer, Thomas, 114
Dwyer, William O., 24

## E

Eagle Mine, xx, 209, 211, 213
Eagle River, xiv, 4, 7
Eagle Rock, 210, 211
Earle, Charles M.W., 18, 24
East Kingsford, Mich., 174
East Saginaw Iron Co., 122, 123
Eddy, Cullen C., 18

233

Eddy, Norman E., 18
Edwards, J.W., 51, 56, 57, 123
Eger, Otto, 105
Elbridge, Mich., 192
Eldorado Silver Mining Co., 15, 27, 34
Ely, Clarence R., 68, 112
Ely, Samuel P., 25, 41, 52
Emery, A.G., 43
Emmet County, 192
Empire Mine, 11, 242
Environmental Protection Agency, 211
Escanaba, Mich., 114, 150, 213
Etten Mining Co., 181
Etten, N., 180, 181
Euclid Gold Mining Co., 153, 155
Eureka Gold Mining Co., 156
Eureka Silver Mining Co., 49
European House, 153
Everett, C.M., 42
Everett, Philo M., 10, 166
Excelsior Silver Lead Co., 24

**F**

Charles T. Fairbairn, 79, 111, 117, 136
Fairbairn, Charles T., 79, 111, 117, 136
Fantine Mining Co., 35, 166
Farnham, C.M., 170
Fay, James S., 109
Fence Lake, 170
Finland, 79, 130
Finney, J., 180
Finney, Walter, 24, 39
Finn, J. Maurice, 117, 118, 120, 121
Finn's Folly. see The Towers
Fire Centre Mining Co., xix, 75, 133, 138, 143, 145
First National Silver Mining Co., 23, 27
Fitzpatrick, Peter, 129
Flaa, James E., 82
Flack, J.O., 136
Flat River, Mich., 192
Flescheim, Joseph, 174
Foard, T.C., 158
Foard, W.A., 158
Fohr, F., 26
Foley, James F., 128, 134, 136
Foley, Joseph C., 96, 134, 171
Foley, Martin, 19
Forbes, Samuel A., 24
Ford, David M., 100
Ford Motor Co., 134, 136
Forestville, Mich., 22
Forster, John H., 32, 34, 166
Fortuna Mining Co., 34
Fort William, 29, 175
Francis A shaft, 161, 163
Fraser, Chalmers and Co., 46, 53, 62

Fredericks, L.C., 104
Frei, Peter F., 158
Frue vanner, 56, 64, 65, 69, 72, 75, 82, 105
Frue, William Bell, 39, 56, 65
Fuller, George, 15, 16

**G**

galena, xvii, xviii, 7, 8, 13, 14, 15, 19, 22, 25, 26, 27, 32, 33, 34, 65, 129, 140, 142, 143, 149, 150, 172, 175, 179, 181, 189, 190, 204, 207
Galena Creek, 178
Galena, Ill., 5
Galena Silver Co., 20, 28
Galena Silver Mining Co., 140, 190
Gannon, Joseph M., 128
Gardiner, William, 98
Gates five-head stamp mill, 47
Gates gyratory crusher, 68
Gay, Edward B., 42
Gaynor, Thomas, 119
Georgetown, Colo., 47
Germany, 45
Ghiselin, George R., 140
Giant Gold and Silver Mining Co., 114
Gibbs City, Mich., 197
Gillett, Shadrach, 9
Gillis, Cornelius, 175
Gillis Creek, 178
Gingrass, Joseph, 104
Gingrass, Peter, 96, 103, 104, 114, 115, 129
Gingrass, Victoria, 114
Gitchie Gumie Gold Mining Co., 113, 123
Glass, James W., 6
Gogebic County, 178
Gogebic Iron Range, xviii, 175, 179
Golconda Mining Co., 34
Goldberg, B.M., 179, 180
Goldberg, L.D., 179, 180, 183
Gold Lake Mining Co., 110, 111
Gold Mine Creek, vi, xv, 132
Gold Mine Lake, xv
Goldmine Truck Trail, 197
Goodland, Walter, 179, 180
Gordon, J.M., 42
Gordon, John R., 139, 170
Gordon Prospect, 139
Gottstein, Peter R., 159
Gourdeau, A.E., 122
Grabower, Louis, 162
Grace, 46
Grace Furnace, 157
Grand Haven, Mich., 192
Grand Rapids and Ishpeming Gold and Silver Co., 123
Grand Rapids, Mich., 111, 123, 200
Grand River, 192
Granite Point, 7

Grant, C.B., 99
Grant Mineral Land Co., 24, 25
graphite, 149, 167, 168, 173
Graveldinger, Michael, 20, 21
Grayling Gold and Silver Mining Co., 118, 119
Grayling, Mich., 118
Great Depression, xvi, 174, 195
Green, Benjamin E., 6
Greenville, 49, 192
Greenwood, 32, 33
Greenwood Furnace, 33
Gregory, E., 9
Griffey, Clinton G., 128
grizzly, 63
Grove, Jack, 197, 198
Grummett, George, 97, 114, 169
Grummett Gold and Silver Mining Co., 169
Grummett Lake. see Ned Lake
Guenther, Richard, 178

**H**

Hall, Edward R., 96, 112, 114
Hall, William F., 20
Halstead, Nathaniel, 44
Hancock, 10, 15, 19, 25, 44, 130, 134, 169, 170
Hancock Iron Mining Co., 169
Hancock Silver Mining Co., 44
Hanna, Robert, 16, 17
Hanscom, Irving D., 122
Hanson, Rasmus, 118
Hardinge ball mill, 106
Harlow, Amos R., 51
Harlow Farms, 152
Harrington, Charles, 183
Harris, William, 15, 19, 28, 39
Hart, Mich., 192
Hartsfeld portable furnace, 184
Harvey, Mich. 60
Haselton, Hyatt S., 143
Hayden, George W., 110
Hayes, John T., 129
Healy, W.P., 113
Hebard, Charles, 44
Heberlein, A., 53
Hematite Mining Co., 136
Henderson, R.R., 184
Henry, Alexander, xvii, 2
Henry, D.F., 58
Henszey, Joseph, 17, 18
Heth, Henry S., 140
Hickok, J.A., 180
High, Sarah F., 140
High, William B., 140
Hill, Samuel Worth, 4, 11, 169
Hodge, Jim, 196
Hodgkins, George, 162
Hodgson, J.C., 10
Hogan, Phillip, 129
Holland, John, 19
Holliday, Jim, 173

Hollinger Mine, 84
Holmes, C.B., 184
Holyoke Mining Co., iv, xvii, 25, 28, 29, 30, 32, 33, 34, 35, 121, 137, 139, 146, 149, 203
Holyoke Trail, 30
Homer, Alaska, 29
Hooper, Thomas, 40, 57, 59
Hopper, W.E., 172
Hornstein, Albert, 152
Horricon Lake. see Ned Lake
Houghton, iv, xiv, xvii, 3, 4, 5, 6, 9, 10, 14, 15, 16, 20, 21, 22, 26, 35, 49, 56, 83, 100, 107, 136, 145, 159, 166, 172, 173, 197
Houghton, Douglass, xiv, xvii, 3, 4, 5, 6, 15, 168
Houghton, Jacob, xvii, 4, 9
Howard City, Mich., 192
Howard, George, 9
Howell, Charles M., 58
Hubbell, Jay, 44, 45
HudBay Minerals Inc., 207
Hudson, Radcliff, 6
Hughes, T., 123
Hugo, Victor, 35
Humbert, Jake, 31
Humble Oil Co., 202
Humboldt, xx, 49, 84, 88, 89, 94, 117, 212, 213
Humphrey, Samuel, 42
*Humphrey's Photographic Journal*, 42
Huntington centrifugal mill, 73, 74, 100, 105, 119
*Hunt's Merchant's Magazine*, 198
Hurley's Huron Mountain Slate and Mining Co., 167
Hurley, Timothy T., 167
Hurley Tribune, 188
Huronian Marble Co., 61, 62
Huron Mountain Exploring and Mining Co., 171
Huron Mountains, 168, 171
Huron Mountain Silver Mining Co., 18
Huron River, 21, 171, 172, 198
Hursley, Burr, 51, 52

**I**

Idaho Mining Co., 24
Ionia County, 192
Iron Bay Foundry, 58, 157
Iron Bay Manufacturing Co., 67
Iron Cliffs Co., 147, 159
Iron County, ix, 173, 192
Iron Mountain, 174
*Iron Ore*, vi
Iron River, xvii, xviii, 2, 36, 38, 41, 42, 43, 44, 45, 46, 47, 48, 49, 50, 51, 52, 54, 56, 57, 58, 59, 104, 114, 140, 151, 167, 192, 196, 197

Iron River Silver Mining Co., 43, 45
Iron River Silver Range, 51
Iron River Silver Reduction Works, xvii, 47
Ironwood, Mich., 179
Isabella Silver Lead Mining Co., 20, 28, 166
Ishpeming,, v, vi, ix, xv, xviii, xix, 11, 32, 38, 43, 49, 53, 60, 62, 69, 76, 78, 79, 82, 83, 84, 96, 97, 100, 101, 104, 105, 107, 108, 109, 111, 114, 117, 119, 120, 121, 122, 123, 124, 129, 130, 132, 133, 134, 135, 138, 140, 141, 142, 143, 146, 147, 148, 150, 151, 157, 158, 165, 168, 170, 175, 183, 192, 202, 242
Ishpeming Gold and Silver Exploring Co., 120, 122
Ishpeming Gold Mines Co., xix, 104, 117
Ishpeming Gold Range, 69, 108, 109, 111, 119, 124, 142, 168, 192, 202
Ishpeming Gold Syndicate, 123, 147
Ishpeming National Bank, 96
Ishpeming Silver Mining Co., 49

**J**

Jackson, Abraham A., 20
Jackson, Charles C., 17
Jackson Silver Mining Co., 20, 28
Janesville, Wis., 141
Jay Gould Mining Co., 190
Jenkins, John, 110
Jenks, James P., 18
Jenney, F.B., 38
Jesuit, 2
Jochim, John W., 129, 130
Johnson, Gilbert D., 53
Johnson, J.M., 43
Johnson, Robert, 78
Johnson, Sven, 80
Johnson, Waldo, 116
Jones and Laughlin Steel Corp., 172
Jones, Byron, 135
Jones, Jacob P., 17, 18
Jones, Will L., 96
Jopling, James, 96, 170
Jopling, James E., 147
J. Ropes & Co., 60
Judd & Crosby Reduction Works, 47

**K**

Kalkaska County, 192
Kaufman, Nathan M., 112, 113
Kaufman Sports Complex, 157
Kennecott Exploration, 210, 211

Kennecott Minerals, xx, 159, 211, 212
Kennedy, Cornelius, 138
Kent County, 192
Kerr McGee co., 203
Keweenaw, 41
Keweenaw Bay, 168, 211
Keweenaw County, 166
Keweenaw Silver Mining Co., 34, 166
Kibbee, Chandler, 21
Kibbee, Henry C., 166
King, Henry B., 106, 107
Kingsford, Mich., 87
Kingston, John T., 175
Kinney, Aaron, 197
Kirk, Wallace, 160
Kirkwood, Charles, 133
Kloeckner, D., 134
Knight, Henry C., 9
Knight, John A., 164
Kobi, Charles, 142
Korten, George, 129
Korten Gold and Silver Co., 129, 197
Krelwitz, Edmund, 49
Krieg, Charles, 163, 164
Krieg, Eugene J., 163
Krieg, Frank E., 162, 163, 164
Krieg, John G., 163
Kylmanen, Abram, 79

**L**

Lady Campbell Gold Mining Co., 120
La Estrella Gold and Silver Mining Co., 176
LaHam, J.G., 107
Lake Gogebic, xviii, 175, 176, 177
Lakeshore Copper Mine, 87
Lakeshore Inc., 87
Lake Superior and Ishpeming Railroad, 148, 161
Lake Superior Gold Mining Co., 122
Lake Superior Iron Co., xviii, 11, 96, 97, 109, 110, 112
Lake Superior Land Co., 17
*Lake Superior Magazine*, 244
Lake Superior Mining Co., xvii, 6, 7, 109
*Lake Superior News and Journal*, 18
Lake Superior Port Cities Inc., ii
Lake Superior Powder Co., 156
Lake Superior Silver and Lead Co., 176, 177
Lake Superior Silver Lead Co., 13, 14, 26, 27, 30
Lane band friction hoist, 69
Langdon, George C., 23
Langdon, John, 15
L'Anse, 9, 34, 167, 168, 171, 173
LaPlata Mining Co., 166

235

Larsen's Gold, 196
Leadville, Colo., 30
Learned, Charles G., 7
Learned, Edward Jr., 6, 7, 30
Leelanau County, 192
Lehnen, N., 188
Leonard, William, 183
Little Alice Gold Mining Co., 120
Little Huron River, 21
Little Iron River, xvii, 36, 38, 43
Little Sable River, 192
Little Traverse River, 192
Livermore, B. Rush, 24, 25
Locke, John, 146
Longyear, John M., vi, 158, 159
Loranger, Joseph, 138
Lord, Henry W., 9
Love, C.L., 123
Lowell, Mich., 192
Low, Thomas, 26
Ludwig, John W., 159
Lundin, Frank, 82, 202
Lundin Mining Corp., 213
Lundquist, Peter J., 135
Luzerne Silver Mining Co., 49
Lyons, W.L., 96

## M

Maas, George, 131
Madden, Phillip, 129
Magoon, Alfred, 116
Malm, John, 82
Malott, Don, 195
Mammoth Silver Mining Co., 43, 44, 151
manganese, 172
Manistee County, 192
Manistee River, 192
Manser, Clarence F., 107
Mapes, E.J., 42
Maple River, 192
Marcellus, Mich., 192
Marchand, Ben, 174
Marcy, William Learned, 6
Marion, Wis., 179
Marius Mining Co., 35, 166
Marks, John, 170
Marlett, William J., 6
Marquette, iv, vi, ix, xi, xiv, xv, xvii, 5, 7, 10, 11, 12, 13, 14, 17, 18, 20, 21, 22, 23, 24, 25, 27, 32, 33, 35, 38, 39, 41, 42, 43, 44, 48, 49, 50, 51, 53, 56, 58, 60, 63, 64, 65, 67, 70, 72, 76, 85, 96, 98, 99, 100, 102, 104, 105, 106, 108, 111, 112, 114, 115, 119, 122, 127, 135, 137, 138, 146, 150, 151, 152, 154, 155, 156, 157, 158, 159, 160, 161, 163, 164, 166, 167, 168, 170, 171, 177, 192, 193, 194, 202, 209
Marquette and Ontonagon Railroad, 24, 33

Marquette County, iv, vi, ix, xi, xv, 10, 12, 13, 14, 18, 20, 21, 22, 24, 25, 44, 53, 60, 63, 64, 65, 70, 72, 76, 85, 102, 135, 137, 138, 146, 156, 160, 161, 166, 192, 194, 209
Marquette County Historical Society, vi
Marquette Gold and Silver Co., 152, 170
Marquette Golf and Country Club, 151
Marquette Iron Range xiv, 11
Marquette Mines Inc., 108
Marquette Mining Co., 17, 24, 35, 104
Marquette *Mining Journal*, 44, 193
Marquette Silver Mining Co., 18, 27, 34, 137, 138
Marsh, Henry B., 9
Martin, Charles P., 25
Martin, John H., 77
Martin, John T., 14, 15
Mason, Stevens T., 3
Masser, George A., 143
Mather, Henry R., 25, 146
Mather, Samuel, 97
Maxwell, Robert, 133
McClear, James L., 130
McCluskey, J.A., 183
McDonald, August, 140
McDonald, John, 114, 140, 141
McEntee, Thomas M., 22
McGhee, George, 172
McGinley, E.H., 169
McGregor, Jason, 24
McKay Co., 194
McKellar, Peter, 175
McLaughlin, Hugh, 174
McLean, James M., 14
McMahon, C.W., 129
Meads, Alexander, 131
Meads, Alfred, 38
Meads, Thomas, 38, 152
Meeker, A.B., 51
M.E. Harrington and Son, 69, 119
Melby, Christian, 138
Mellen, Thomas, 52, 54, 57
Menominee, 49, 173, 174, 206
Menominee iron range, 173
Menominee River, 174, 206
*Merchant*, 9
mercury, 46, 47, 48, 54, 58, 64, 65, 67, 74, 75, 82, 106, 188
Merritt, Daniel, 58, 156, 157
Merryweather, Henry, 17
Metric Gold and Silver Co., 188
Mexican American War, 25
Michelson, Nels, 118, 120
Michigamme, Mich., 168, 169
Michigan College of Mines, 172

Michigan College of Mining and Technology, 83, 106, 107, 194, see also Michigan Technological University and Michigan College of Mines
Michigan Copper Mining Co., 79
Michigan Department of Environmental Quality, 211
Michigan Department of Natural Resources, 203, 210, 211
Michigan Geological Survey, vi, 38, 150, 192, 199
Michigan Gold Co., 97, 99, 100, 101, 102, 103, 111, 112
Michigan Gold Mine, xi, xvi, xviii, 74, 96, 97, 107, 108, 123, 134, 202
Michigan Gold Mines Inc., xix, 106
Michigan Gold Mining Co., xix, 107, 203
Michigan Gold Range, 108, 110, 122, 142
Michigan Iron Co., 33
Michigan Land and Iron Co., 129, 133, 134, 136, 140, 141, 142
Michigan Quartz Silica Co., xix, 106, 112
Michigan State Highway Department, 195
Michigan Technological University, 86, 203, 210
Middle Bay, 5, 8
Migisy Bluff, 158
Mildon, Henry H., 113, 134
Miller, John, 150
Mills, Frank P., 97
Milwaukee, 58, 87, 106, 107, 123, 126, 143, 175, 179, 186
Milwaukee and Lake Superior Silver Mining Co., 58
Milwaukee, Lake Superior and Western Railway, 179
Milwaukee, Lake Superior and Western Railway, 186
Minerals Processing Corp., 108, 203, 206, 207
Miner, Anson B., 96, 112, 114, 134, 135, 136, 169
Miner, George B., 175
Minesota copper mine, 15, 28, 39
*Mining and Engineering World*, 172
Minneapolis, 123, 181, 182, 183, 184, 187, 188, 189
Minneapolis and Gogebic Mining Co., 188
Miracle, George, 180, 183, 186
Mitchell, Peter, 177
Mitchell, Samuel L., 100, 101
Mockler, John, 121
Mockler, Richard, 121
Mohatin Hills Mine, 172
Mohawk Mining Co., 18, 34, 49
Molloy, James, 114, 140

236

Monitor Mining Co., 17, 24
Montcalm County, 192
Montreal River, 6, 7, 179
Montrose, Mich., 200
Moore, Francis M., 163
Moore, Smith, 152, 153, 154, 155
Moore, Terrence, 39, 41, 43, 46, 151, 152, 171
Moore, W.A., 166
Moran, Frederick T., 116
Moran, William B., 116
Moreau, Samuel, 178, 180
Morgan Creek, 148
Morgan Heights Sanatorium, 148
Morgan Iron Co., 24
Morgan, James, 131
Morgan, Mich., 24, 131, 147, 148, 244
Morris, Angus, 198
Morse, Jay C., 51
Moss, Joseph L., 17
Mountain Beauty Gold Mining Co., 120
Mount Mesnard, 158
Muck, Charles, 148
Murdoch House Hotel, 133
Murdoch, William, 133
Murray, David, 43
Muskegon River, 192
Myers, A.W., 101
*Mystic*, 23, 160

**N**

National Co., xvii, 9, 10
National Mining Co., 9
Ned Lake, 169
Negaunee, ix, xiv, xv, xvii, 4, 12, 32, 43, 49, 101, 113, 118, 123, 127, 128, 129, 130, 131, 132, 134, 135, 136, 140, 146, 147, 150, 151, 196, 242, 243
Negaunee Gold and Silver Co., 128
Negaunee Gold and Silver Mining Co., 130, 131
Negaunee Mining Co., 49
Nelson, E.D., 104
Nelson, Robert, 113
Nester, Timothy, 170
Nevada Mining Co., 22
Newaygo County, 192
Newbury, Ver., 60
Newport, Ken., 183
Newton Mining Co., 183, 185
New York and Lake Superior Mining Co., xvii, 6, 7, 30
New York and Michigan Silver Lead Co., 34
New York Mining Co., 49
*The New York Times*, 200
Nichols, A.F., 107, 184
Nichols, E.M., 184
Nicor Mineral, 202

Niemi, Abel, 82
Nomex Au Joint Venture, 202
Nonesuch copper mine, 59
Noon, William, 104
Norgan Gold Mining Co., 136, 145
North American Mining Co., 11
North Beaver shaft, 145
*Northern Light*, 16
Northern Light Silver Mining Co., 16, 19, 24
Northern Lumber Co., 164
Northern Michigan University, 242
Northrop, George, 43
Northrup, C.S., 181, 183, 185
Northrup, J.A., 181, 183
Northrup, Walter A., 19, 20, 21, 28, 29
North Silver Lake Mining Co., 16, 22, 26
*North Star*, 16, 17
North Star Silver Mining Co., 16, 19, 24
North Washburn Mining Co., 187, 188
Norway, Mich., 173
Nosack, F., 180

**O**

Oakland County, 192, 195
Oat, George R., 18
Oceana County, 192
Oconto, Wis., 140
O'Grady, Bernard, 16
Oja, Henry, 79
Ojibwe, 2, 5, 34, 36, 173, 200
Old Veteran Mining Co., 184, 190
Oleson, Ole, 157
Olmstead, Albert F., 183
Ontonagon, 39, 41, 44, 47, 56, 58
Ontonagon and Lake Superior Silver Mining Co., 42, 51, 56
Ontonagon and State Line Railroad, 37
Ontonagon Central Silver Mining Co., 42
Ontonagon County, iv, xi, xv, 36, 42, 151, 167, 175, 192, 196
Ontonagon, Mich., iv, xi, xv, xvii, 2, 3, 9, 24, 33, 36, 37, 38, 39, 40, 41, 42, 43, 44, 45, 46, 47, 49, 50, 51, 52, 53, 54, 55, 56, 58, 151, 167, 175, 177, 178, 192, 196
Ontonagon Miner, 38
Ontonagon River, 2
Ontonagon Silver Mining Co., 39, 40, 41, 44, 45, 55, 151
Ontonagon Silver Range, 37
Original Sauk's Head Mine Ltd., 162, 193
Orr, J. Forrest, 81
Orr, Wesley B., 106
Ortonville, Mich., 200
Osage Mining Co., 18, 34, 35

Osborne, Idaho, 86
Oshkosh, Wisconsin,, 177, 178
Otsego Mining Co., 17, 24
Ottawa County, 192
Otter Head Tin Swindle, 28
Otter Lake, 10

**P**

Paint River, 197
palladium, xiv, xx, 211, 213
Palmer, Leander C., 24
Palmer, Nathaniel B., 14
Palmer, Waterman, xvii, 11, 60
Palms, Francis, 115, 116
Panic of 1873, 45, 57
Panic of 1893, 71, 103, 145
Parmlee, George, 114
Parsons, Edwin, 146
Paul, Joe, 84
Paulson, John, 123
Payne, H. J., 114
Peninsular Gold Mining Co., 116, 118, 135, 139, 192, 203
Pennock, Homer, 17, 24, 27, 28, 29
Pennock, Prosper, 43
Pepin, Joseph, 120
Pepin, Leon, 122
Perch Lake, 173
Pere Marquette Mining Co., 104
Perry, Mich., 200
Peshekee River, 198
Peter Paul's Gold Mine, 197
Peterson, Charles, 132
Pfister, Charles F., 143
*Phantom*, 41
Phelps Dodgem co., 203
Philadelphia, 11, 17, 18, 19, 25, 63
Phillips Gold Mine, 124
Phillips Gold Mining Co., 124, 126
Phillips, John, 124
Phoenix, Ariz., 85
Phoenix copper mine, 84
Pickands, James, 51, 97
Pickands, Mather and Co., 97
Pierpoint, James, 50
Pine Creek, 173
Pings, Andrew S., 158
Pittsburgh, 11, 43, 50, 51, 98, 151, 194
Pittsburgh and Lake Superior Iron Co., 11
Pittsburgh Silver Co., 50
Placer Mining, 191
platinum, xx, 211, 213
plumbago, 173
Porcupine Mountain, 59
Porcupine, Ont., 84
Porcupine Silver Mining Co., 49
Portage Lake, 10, 15, 25, 34, 166
Portage Lake Silver Mining Co., 166
Portage Lake Smelting Works, 15, 25

Pratt, C.H., 48
Pratt, W.A., 12
Presque Isle, 7, 8, 36
Presque Isle River, 36
Primeau, Peter, 162
Prospectors Creek, 178
Purdon, George R., 163
pyrite, 7, 19, 65, 75, 89, 90, 129, 132, 133, 142, 143, 151, 153, 172, 174, 177, 179, 181, 192, 204, 207
pyrrhotite, xiv, 211

**Q**

Quinnesec, Mich., 174

**R**

Ralston, Charles, 126
Ralston, James H., 15
Rand Drill Co., 80
Rand "Little Giant" drills, 66
Rapid River, 192
Rasmussen, George, 193
Ratigan, W.P., 116
Raymond, Albert, 114
Raymond, George, 114
Ray, W.J., 168
Reany Lake, 202
Rees, Allen F., 136
Rensselaer School, 3
Republic iron mine, 84, 152, 165
Republic, Mich., 165
Republic Steel, 84
Resource Exploration Inc., 85
Richards, Joseph M., 16
Richardson, Alex, 141
Ripka, A.A., 51
Ripley, Mich., 85
Rivière Ste. Anne. see Iron River
Robbins, Byron P., 43
Robbins, Edward, 132, 135
Robinson, Dean, 136
Robinson, Orrin W., 10, 173
Rocking Chair Lakes, 137
Rock Kilns, 75
Rockland, Mich. 19, 79
Roessler, Christopher, 130
Rogers, Benjamin T., 16, 19
Rogers, George W., 9
Rominger, Carl, 150
Rood, Anson H., 18
Rood, William H., 81, 102, 124
Roosevelt, Teddy, 121
Ropes Gold and Silver Co., xviii, xix, 62, 68, 73, 75, 99, 121, 126
Ropes Gold Mine, 60, 134, 138, 141, 202, 208, 212
Ropes Gold Mine Public School, 76, 77
Ropes Gold Range, 125, 142
Ropes, Julius, xviii, xix, 11, 38, 60, 61, 81, 121, 126, 142, 151, 155, 174, 183, 193

Ropes, Leverett, 81
Ross, James B., 15, 22
Rotunda Building, 20, 22, 23
Rouse, Eunice Louisa, see Smith, Eunice Louisa
Royal Oak Gold Mining Co., 120
Roy, Theophile, 130
Ruppe, Charles, 138, 142
Russell, George, 105
Ryan, Edward, 10
Ryan, John F., 171

**S**

Salmon Trout River, 18, 21, 211
Sands Station, 150
Sanger, Henry K., 18
Sanger, Henry P., 18, 166
Sanson, John, 115, 139
Sault Ste. Marie, 8, 9
Saux Head, 161, 163, 164
Saux Head Copper Mining and Development Co. Ltd., 163, 164
Sawmill Creek, 194
Schonnaway, Fred, 173
Schoolcraft, Henry Rowe, 3
Schroeder, A.F., 107
Schroeder, Walter L., 107
Schweisenthal, Mathias, 128
Scranton, Penn., 44
Scranton Silver Mining Co., 43, 44, 45, 53, 55
Secor, Cora, 106
Section Twenty Five Co., 190
Albert K. Sedgwick, 79, 117
Sedgwick Mining Co., 146, 147
Selden, Richard L., 150
Sellwood, Joseph, 124, 134, 136
Shafer, Jacob, 21
Shelden, Ransom, 15, 18, 19
Sherman, Watts, 6
Shiawassee County, 200
Shiawassee River, 200
Shuneaw Silver Mining Co., 34
Sibley, Alexander H., 13, 14, 30
Sibley, Frederick B., 13
Sibley, Henry Hastings, 14
Sibley Peninsula, Ont., 65
Sibley, Solomon, 14
Silver City, Mich. xv, 48, 49, 196
Silver Creek, 141, 203
Silver Islet, 29, 30, 39, 41, 56, 65
Silver King Mining Co., 183, 185
Silver Lake, xv, xvii, 13, 14, 15, 16, 19, 21, 22, 25, 26, 30, 139, 142, 151
Silver Lake Mining Co., 22
Silver Lake Trail, 21, 30
Silver Lead Creek, 150
Silver Lead Mine Lake, xv
Silver Mine Lake, xv, 143
Silver Mining Co., xviii, 15, 16, 18, 20, 21, 23, 27, 30
Silver Mountain, xv, xvii, 9, 10, 49

Silver Mountain Mining Co., 49
Silver River Graphite Co., 167
Sioux Mining Co., 34, 166
Slockett, Henry, 30, 32
Smith, David, 14
Smith, Eunice Louisa, 60
Smith, Gad, 150
Smith, H.N., 58
Smith, L.C., 9
Smith, Silas C., xvii, 12, 14, 146, 147, 151
South Beaver shaft, 144
South Shore Silver Mining Co., 43
South Washburn Mining and Smelting Co., 184, 189
Spaulding, John, 16, 17, 19, 24, 44, 50
Spearbracker, C.C., 180
Spear, John W., 155
Spencer, George, 160
Spiroff, Kiril, 107
Sporley, Gottlieb, 148
Springfield, Ill., 124
Stafford, Henry H., 25, 60, 155
St. Clair, 56
Stevens, J.F., 128
Stewart, N.H., 111
St. Joe American Corp., 203
St. Johnsbury Academy, 60
St. Joseph County, 192
St. Lawrence iron mine, 75
St. Louis River, 5
Stone, J.W., 80
Strathman, Elmer, 134-5
Stringer, Charles A., 163
Strobeck, D.F., 181
Sturgeon River, 10, 173
Sucker Creek, 129
Summit Exploration, Mining and Manufacturing Co., 177
Sunday Lake, 175, 178, 179
Suneson, Martin, 165
Superior Gold and Silver Co., 111, 112, 113, 126, 127, 142, 203
Superior Oil Co., 202
Superior Silver Mining Co., 41, 46, 48, 151
Swain, Hosea B., 115
Swallow, 8
Sweet, Bemsley G., 16, 19, 24
Swineford, Alfred P., 43, 44

**T**

Talcott, Andrew, 6
Talcott, George, 6
Talcott Harbor, 5, 6, 8
Talcott, Sebastian Visscher, 6, 7
Taylor, R.H., 135
Teal Lake, iv, 12, 129, 136, 196, 197
Teal Lake Gold and Silver Exploring Co., 129
Tebo, John, 161
Tecumseh, Mich., 198

Ten Kilns, 133, 138
tetrahedrite, 65
Thayer, L.O., 183
Thayer, O.S., 183
The Towers, 121
Thomas, Edmund, 85
Thompson, John, 171
Thompson, Montgomery, 138
Thunder Bay Gold Corp., 199
Thurber, Henry C., 44, 127
Tilden Mine, 242
Tillson, F.P., 60
Tindall, Burnell, 195
Tislov, W.O., 129
Tobin, James, 176, 178
*Tom Boy*, 160
Toutloff, Moses B., 123
Tower, F.D., 162
Trakt, August, 79
Traverse, Richard P., 39, 42
Treaty of La Pointe, 5
Trebilcock, Albert, 82
Trebilcock, James, 82
Trebilcock, William, 82
Trevarthen, Richard, 114
Trevithic, Thomas, 103, 116
Tribullion Mining, Smelting and Development Co., xix
Tribullion Mining, Smelting, and Development Co., 104
Trombley, Frank, 106
Trotter, Charles W., 18
Trotter, Edward H., 17
Trotter, George, 25
Trotter, William H., 18
Troy, N.Y., 3, 5
Truax, John G., 50
Turner, John E., 19
Turnquist, Frank G., 182
Turnquist, John A., 182

## U

Uncle Billy, see Murdoch, William
Union Gold Co., 21, 28
Upson, Wis., 179
uranium, 172, 200
Urban House, 96
Uren, William, 101

U.S. Geological Survey, 10, 146, 195

## V

Valjean Mining Co., 35, 166
vanadium, 172
Van Buren, Martin, 6
Vandeventer, J.V., 112
Van Dyck, H.H., 6
Van Fleet, Charles, 44
Van Iderstine, Jeremiah, 24
VanKirk, J., 141
Varcoe, Henry, 101
Varcoe, William, 101
Varney amalgamating pans, 47
Vasseur, Louis, 178
Vaughn, R.D., 113, 133
Vernon, Mich., 200
Vertin, Joseph, 162
Victoria, Mich., 114, 192
volcanogenic massive sulfide, 207

## W

*Wakefield Bulletin*, 179, 180
Wakefield, George M., 177, 178, 179
Wakefield Gold and Silver Co., 104, 183, 186
Wakefield, Mich., 179, 180
Walker, Henry W., 140
Walker, H.O., 122
Wallace Diamond Mining Inc., 86
Walton, Mich., 192
Wanzer, Albert, 15, 19
Ward, William, 98, 115
Warner, W.W., 180, 181, 183, 184, 186, 188, 189, 190
War of 1812, 25
Washburn, G.W., 181
Washburn Mining and Milling Co., 187, 189
Washburn Mining Co., 181, 184, 187
Washtenaw County, 195
Watervliet, N.Y., 5, 6
Weatherston, George, 66, 67, 102, 142
Weber, William C., 198
Weeks, James A., 9

Wells, Daniel Jr., 175
Wells, Thomas M., 128
Wentworth, Frank B., 75
Westbrook, W.E., 50, 51
Westby, A.D., 181, 182, 184
Western Mining Corp., 203
West Summit, Mich., 192
Wetmore, Charles H., 18, 32
Wetmore, W.L., 50
Wexford County, 192
Whetstone Brook, 12, 151
Whitehall, Mich., 192
White, Peter, ix, 25, 96, 97, 99, 100, 103, 105, 111, 112, 147, 152, 153, 154, 170
White Pine Mine, 59
White River, 192
Whitney, W.M., 181
Wilkinson, James, 159
Williams, Charles P., 11
Williams, James, 43
Williams, John, 15, 25
Williams, Mose, 33, 34
Williams, Nathan, 23
Windom Gold and Silver Mining Co., 183, 188
Winthrop iron mine, 123
Wisconsin Geological Survey, 51
Wiswell Electric Ore Pulverizer, 73, 74, 154, 168, 188
Wolverine Mining Co., 49
Woods, J.H., 9
World War I, 59
World War II, xvi
Wright, Austin W., 140
Wright, Charles E., 51
Wright, E.L., 10
Wyandotte Silver Smelting and Refining Works, 41, 56

## X

xanthate, 89

## Y

Yellow Dog Plains, 194, 195, 209
Yellow Dog River, 194, 195, 198
Yellowstone Mining Co., 17, 24
York, Samuel, 161

## Z

Zhulkie Creek, 127

239

# Index of Graphics

1887 mill at the Ropes Gold Mine, 19

**A**
A.B. Meeker & Co. ad, 49
A.W. Myers & Co. ad, 100
Abram Boulson tailor ad, 130
Adams, John Q., 128
Alger, Russell, 129
Anthony, Edward, 42
Arthur Tailor ad, 133

**B**
B.F. Childs' photography ad, 39
Ball & Ball, attorneys ad, 155
Ball, Dan, 155
Beaver Board adit, 149
Begole, Frederick, 133
Big Bay ad with map, 193
Blake's Stone and Ore Breaker, 40
Breitung, Edward, 148
Buell, Jonathan, 174

**C**
C. Melby & Co. ad, 138
C.H. Kirkwood & Robert Maxwell ad, 132
C.M.W. Earle ad, 22
Callahan's primary grinding mill, 84
Camp Gray at Talcott Harbor, 6
Capt. M. Daniel ad, 160
Charles Kobi tailor ad, 142
Charles Wright ad, 50
Chippewa Mining Co. mine, 167
Cleaves, W.S. 172
Cooper, James, 101
Cornelius Kennedy justice of the peace ad, 140
Corser, Austin, 58
Courtis, W.M., 191
Cresent adit, 143
Cummings, George, 21
Curry shaft house, Ely shaft, Ropes 1887 mill building, 63
Curry, Solomon, 62

**D**
Dead River Gold Range, 137
Dear Lake furnace, 126
Detective Mining Co. ad, 182
Detroit Gold and Silver Co. adit, 141
Drawing, modern mining, Ropes, 87
Drilling blast holes, Ropes Gold Mine, 78
Drum disc and filters, 83
Dunlap Mining Co. ad, 185

**E**
E.R. Ely broker ad, 112
Eagle decline drilling, 210
Eagle Mine diagram, 209
Euclid Gold Mining Co. stock, 155
European Hotel, 154

**F**
F. Braastad & Co. ad, 111
Finn, J. Maurice, 117
Flotation cells and cyanide leaching tanks, 86
Flotation cells at Callahan's Humboldt Mill, 20
Forster, John, 32
Fountain, Dan, 243
Frank Krieg sketch of family gold mine, 162
Frue vanner in Ropes 1884 mill, 64

**G**
Gad Smith, 150
George Cummings ad, 120, 146
George Wakefield mining ad, 188
Gogebic Gold Range, 176
Gordon prospect adit, 139
Grand Rapids and Ishpeming Gold and Silver Mining Co. stock, 120
Grayling Gold & Silver Mining Co. stock, 118
Grummett Gold and Silver Mining Co. stock, 171

Grummett, George, 114

**H**
H.H. Stafford ad, 147
Hand-powered blower, 30
Headframe, circa 1982, 93
Hebard, Charles, 44
Heron Mountain Silver Mining Co. stock, 23
Hiram Burt real estate ad, 41
Holyoke shaft, 28
Hooper, Thomas, 57
Horse whim, 26
Hotel Toutloff ad, 122
Houghton, Douglass, 3
Hubbell, Jay, 45
Huntington's Centrifugal Roller Quartz Mill ad, 103
Huron Bay Slate and Iron Co. ad, 168

**I**
Iron Bay Manufacturing Co. ad, 159
Iron River Reduction Works, 52

**J**
J.C. Foley insurance, real estate ad, 98
J.M. Longyear ad, 158
J.W. Spear ad, 156
James McClear attorney ad, 130
James Pickands & Co. ad, 46
Jochim, John, 130
Julian M. Case ad, 157

**K**
Kaufman, Nathan, 112
Kingston, John T., 175
Krieg's Gold Mine, 164
Krieg's mine crew, 165

**L**
Lake Superior Iron Co. ad, 109
Lake Superior Silver Lead Co. prospectus, 33

Lake Superior Silver Lead Co. stock, 34
Livermore, R. Rush, 24
Longyear, John M., 158
Louis Grawbower ad, 163
Lundin, Frank, 82

**M**
Marquette map circa 1844, 5
Merritt, Dan, 156
Michelson, Nels, 118
Michigan Gold Co. ad, 104
Michigan Gold Company stock, 106
Michigan Gold Mine crew, circa 1935, 105, 107 & 108
Michigan Gold Mine plans, section, 97
Michigan Gold Mine, circa 1905, 102
Michigan Gold Mining Co. stock, 101
Michigan Gold Range, 110
Miner, Anson, 96
Miner's camp at Holyoke, 29
Mitchell, Peter, 177
Mockler Bros. ad, 122

**N**
Negaunee Gold and Silver Mining Co. stock, 131
*North Star*, 16

**O**
Office on edge of Ropes cave-in, 90
Ontonagon Silver Mining Co. stock, 55
Ontonagon Silver Range, 37
Original Sauk's Head Mine Limited, 161

**P**
Palms, Francis, 115
Peninsular Gold Mining Co. ad, 116
Peninsular Mine adit, 232
Peninsular mine diamond-drill work, 116
Penncok, Homer, 27
Peter White agency ad, 99
Phillip J. Hogan beverages ad, 128
Phillips Gold Mining Co. 127

**R**
Richard Blake ad, 123
Robert Maxwell ad, 133
Robinson, Orrin 173
Robinson, Orrin W., 10
Ropes 1887 mill building, 65
Ropes 1937 underground workings chart, 85
Ropes Gold Mine detail, 71
Ropes Gold Mine, circa 1935, 80
Ropes Gold Mine, circa 1955, 88
Ropes Gold Range, 125,
Ropes mills overview, 70
Ropes mine and cyanide plant, circa 1900, 74
Ropes mining crew, 69
Ropes mining crew, circa 1935, 81
Ropes Production Chart, 1883-1897, 95
Ropes public school, 77
Ropes school house, miners' cabins, 76
Ropes stamp mill and amalgamating plates, 67
Ropes vanner room, 72
Ropes, Julius, 61
Rothschild & Bending ad, 157
Rotunda Building in Detroit, 20
Ruppe shaft, 144

**S**
Samuel E. Byrne abstracts ad, 151
Sheldon, Ransom, 18
Sibley, Alexander, 14
Silver lead range circa 1864, 13
*Smith Moore*, 152
Smith Moore's mine, 153
Spaulding, John, 15
Strathman, Elmer, at B&M prospect, 135
Superior Gold and Silver Co. stock, 113
Swineford, Alfred, 43

**T**
T. Hughes ad, 122
T. Meads ad, 38
The Towers (Finn's Folly), 121
Thunder Bay Gold Corporation stock, 199

**U**
Underground loader, 94
Upper Peninsula map, xii-xiii
Urban House ad, 98

**V**
Vasseur, Louis, 178

**W**
W.W. Warner & Co. real estate ad, 180
Wakefield, George, 179
Washburn Mining Co. ad, 187
Weatherston, George, 66
White, Peter, 99

# About the Author

Daniel R. Fountain is a historian with a deeply rooted appreciation for Michigan's mining heritage and an avid interest in the rich and colorful history of the early settlers and miners.

Dan grew up in the Upper Peninsula mining town of Ishpeming, surrounded by the open pits and towering headframes of the area's iron mines. Much of his understanding and curiosity about the region's mining history – and his own interest in searching out old mining sites – stems from growing up in that atmosphere.

Three generations of his family were involved in the iron industry, so it was no surprise that Dan made his living in the mines. He worked more than 30 years at the Empire and Tilden mines as an electronic technician. He is a graduate of Northern Michigan University, holding a degree in electronics.

For more than three decades, Dan has conducted extensive research on gold and silver mining in Michigan. He has used the modern tools afforded by digital libraries and online searching, but also has spent countless hours in museums and libraries seeking long-forgotten historical documentation.

To get a better feel for the stories, he has gone out and hiked many miles in the rugged forests of Upper Michigan in search of the numerous gold and silver mining prospects about which he writes.

Dan has served on the board of directors and as president of the Negaunee Historical Society, and continues to be involved in on-going research concerning gold and silver mining, maritime history and other subjects of local historical significance.

In addition to prospecting, in his leisure time Dan enjoys traveling, photography, scuba diving and shipwreck research. Dan and his wife, Judy, live in Negaunee, Michigan.

JUDY FOUNTAIN

# From Lake Superior Port Cities Inc.

**Lake Superior Magazine**
A bimonthly, regional publication covering the shores along Michigan, Minnesota, Wisconsin and Ontario

**Lake Superior Travel Guide**
An annually updated mile-by-mile guide

**Lake Superior, The Ultimate Guide to the Region – Second Edition**
Softcover: ISBN 978-0-942235-97-5

Hugh E. Bishop:
**The Night the Fitz Went Down**
Softcover: ISBN 978-0-942235-37-1

**By Water and Rail: A History of Lake County, Minnesota**
Hardcover: ISBN 978-0-942235-48-7
Softcover: ISBN 978-0-942235-42-5

**Haunted Lake Superior**
Softcover: ISBN 978-0-942235-55-5

**Haunted Minnesota**
Softcover: ISBN 978-0-942235-71-5

Beryl Singleton Bissell:
**A View of the Lake**
Softcover: ISBN 978-0-942235-74-6

Bonnie Dahl:
**Bonnie Dahl's Superior Way, Fourth Edition**
Softcover: ISBN 978-0-942235-92-0

Joy Morgan Dey, Nikki Johnson:
**Agate: What Good Is a Moose?**
Hardcover: ISBN 978-0-942235-73-9

Daniel R. Fountain:
**Michigan Gold & Silver, Mining in the Upper Peninsula**
Softcover: ISBN 978-1-938229-16-9

Chuck Frederick:
**Spirit of the Lights**
Softcover: ISBN 978-0-942235-11-1

Marvin G. Lamppa:
**Minnesota's Iron Country**
Softcover: ISBN 978-0-942235-56-2

Daniel Lenihan:
**Shipwrecks of Isle Royale National Park**
Softcover: ISBN 978-0-942235-18-0

Betty Lessard:
**Betty's Pies Favorite Recipes**
Softcover: ISBN 978-0-942235-50-0

Mike Link & Kate Crowley:
**Going Full Circle: A 1,555-mile Walk Around the World's Largest Lake**
Softcover: ISBN 978-0-942235-23-4

James R. Marshall:
**Shipwrecks of Lake Superior, Second Edition**
Softcover: ISBN 978-0-942235-67-8

**Lake Superior Journal: Views from the Bridge**
Softcover: ISBN 978-0-942235-40-1

Kathy Rice:
**The Pie Place Cookbook: Food & Stories Seasoned by the North Shore**
Softcover: ISBN 978-1-938229-04-6

Howard Sivertson
**Driftwood: Stories Picked Up Along the Shore**
Hardcover: ISBN 978-0-942235-91-3

**Schooners, Skiffs & Steamships: Stories along Lake Superior's Water Trails**
Hardcover: ISBN 978-0-942235-51-7

**Tales of the Old North Shore**
Hardcover: ISBN 978-0-942235-29-6

**The Illustrated Voyageur**
Hardcover: ISBN 978-0-942235-43-2

**Once Upon an Isle: The Story of Fishing Families on Isle Royale**
Hardcover: ISBN 978-0-962436-93-2

Frederick Stonehouse:
**Wreck Ashore: United States Life-Saving Service, Legendary Heroes of the Great Lakes**
Softcover: ISBN 978-0-942235-58-6

**Shipwreck of the Mesquite**
Softcover: ISBN 978-0-942235-10-4

**Haunted Lakes** (the original)
Softcover: ISBN 978-0-942235-30-2

**Haunted Lakes II**
Softcover: ISBN 978-0-942235-39-5

**Haunted Lake Michigan**
Softcover: ISBN 978-0-942235-72-2

**Haunted Lake Huron**
Softcover: ISBN 978-0-942235-79-1

Julius F. Wolff Jr.:
**Julius F. Wolff Jr.'s Lake Superior Shipwrecks**
Hardcover: ISBN 978-0-942235-02-9
Softcover: ISBN 978-0-942235-01-2

www.LakeSuperior.com
1-888-BIG LAKE (888-244-5253)
Outlet Store: 310 E. Superior St., Duluth, MN 55802